JUTE—
FIBRE TO YARN

JUTE
Fibre to Yarn

R. R. ATKINSON, A.T.I.

1965

CHEMICAL PUBLISHING CO., INC.

212 FIFTH AVENUE NEW YORK, N. Y.

Jute - Fibre to Yarn

ISBN: 978-0-8206-0094-9

Chemical Publishing Company:
www.chemical-publishing.com
www.chemicalpublishing.net

First American Edition:

© **Chemical Publishing Company, Inc.** - New York 1964

Second Impression:

Chemical Publishing Company, Inc. - 2011

Printed in the United States of America

Contents

Preface

THIS book is intended mainly for students but it is hoped that it will also be of interest and value to those engaged upon the technical side of the industry. Briefly, the intention behind the book is to give an appreciation of the more important aspects of the raw material, to show the basic principles involved in converting the raw material into yarn, and to demonstrate how the machinery does this.

I am deeply indebted to Dr. H. P. Stout, Director of the British Jute Trade Research Association for his continued interest and assistance, and to the Council of the Association for permission to draw from their Research Reports. My thanks are also due to Mr. P. G. Anderson and Mr. G. C. Stevenson, who were kind enough to criticize the text in a most helpful and constructive manner; to Messrs. James Mackie and Sons Ltd, to Messrs. Fairbairn Lawson Ltd, and Messrs. Giddings-Lewis and Fraser Ltd for their ready assistance in providing technical data and photographs of their machinery. Help with proof-reading and the preparation of the index was given by my wife. For this and for her constant encouragement while the book was being written, I thank her.

Dundee, 1964.

R. R. ATKINSON

The Place of Jute in World Textiles

JUTE is second only to cotton in the world's production of textile fibres. It is estimated that in 1960 about 31 thousand million pounds of fibres were processed throughout the world, cotton accounting for roughly half that quantity and jute following with a consumption of nearly five thousand million pounds. Table I shows the relative proportions of the principal fibres used in recent years compared with the pre-war period. The most interesting feature of the Table is the decline in importance of some fibres and the growth of others as economic and technological changes take place.

TABLE I. TEXTILE FIBRE PRODUCTION†
(Individual figures are percentage of total)

	Average 1934–39	1957–58	1958–59	1959–60
Cotton	57·2	47·6	48·5	49·3
Jute and allied fibres	16·5	16·2	18·1	15·6
Wool (apparel)	7·4	7·0	7·2	7·1
Wool (carpet)	1·5	1·3	1·3	1·3
Rayon (filament)	4·8	7·2	6·2	6·7
Rayon (staple)	2·2	9·0	7·8	8·2
Other man-made fibres	—	3·0	3·0	3·9
Silk	0·5	0·2	0·2	0·2
Flax	1·0	0·8	0·7	0·6
Hemp	8·9	7·6	7·2	7·1
Total (million lb)	20,219	28,482	29,326	30,901

† Sino–Soviet bloc excluded.

The major sources of supply of jute lie within the Commonwealth, chiefly in India and East Pakistan. When the Indian sub-continent was partitioned in August, 1947, the main jute-growing area, East Bengal, was awarded to the newly created state of Pakistan while about three-quarters of the manufacturing capacity fell within the boundaries of the Indian Union. Thus at that time Pakistan had ample supplies of fibre but few mills while India had more mills than she had fibre for. Each country began to make the necessary alterations to its economy, Pakistan developing Chittagong and Chalna, its ports on the Bay of Bengal, so that she could export her raw fibre more easily while, at the same time, every effort was made to set up new mills. India, on the other hand, expanded her acreage under jute cultivation to supply her mills, the export of jute cloth being a powerful currency earner and playing a vital part in the economy of the country.

Jute is grown on a large number of peasant smallholdings and it is rather difficult to arrive at an exact figure for the total acreage but it is estimated that since 1955 about 3,600,000 acres each year have been used for jute growing throughout the world, India and Pakistan between them accounting for some 3,000,000 acres. Jute is also grown in Burma, Formosa, China, Brazil, and Nepal, but at present their production is negligible compared with that of the sub-continent. Fibres allied to jute, such as kenaf and Congo jute, are grown in India, Thailand, and the Congo but again output is comparatively small.

In recent years the world's production of jute and its allied fibres has been running at a level of between 2 and 2·7 million tons annually, true jute accounting for about 80 per cent of this. Though some of the minor growing countries are trying to increase their output, one of the difficulties about successful jute growing on a commercial scale is that plentiful supplies of both water and labour are required. From time to time various types of mechanical harvester have been tried but none are, at the moment, capable of handling the large quantities involved.

The largest centre of the jute industry is the Calcutta area of India where some 70,000 looms produce about 1·25 million tons of jute goods annually. Pakistan follows next in importance with an annual output of some 250,000 tons which, it is planned, will increase to 360,000 tons by 1965. After these two countries, the United Kingdom has the largest industry, capable of producing about 160,000 tons of jute goods each year. The manufacturing emphasis in the U.K. differs from that in India and Pakistan; these last mentioned countries being mainly concerned with weaving cloth for sacking and bagging. In Great Britain about one-third of the output is yarn for the carpet industry, and the weaving of speciality fabrics is carried on in preference to sacking fabrics. The combined 'Common Market' countries process some 280,000 tons of jute annually and here again the emphasis is on jute for special purposes. Other countries such as Brazil, Japan, and the United States have smaller manufacturing capacities used mainly for internal trade. India is the world's major exporter of jute cloth, sending large quantities to America for baling cotton and to Australia for grain and wool packing.

Jute has long been recognized as a cheap, strong, durable fabric eminently suited for sacks and bags and many other purposes. On a world basis about 80 per cent of all the jute manufactured finds its way into packing of one sort or another. The actual weight of jute used per ton of transportable material depends on local variations in sack dimensions, whether the goods are for export, whether the bag is returnable or not and so on, but typical figures for the weight of jute used to pack one ton of various products are

Flour	(hessian bags)	15 lb
Flour	(twill sacks)	41 lb
Potatoes	(hessian bags)	22 lb
Potatoes	(twill sacks)	46 lb
Beet pulp		24 lb

In certain cases the contents of the bag must be protected against contamination by the jute itself, by other products stored nearby, or by the atmosphere. For such uses the bag may be lined with paper or polythene bonded to the jute. Alternatively a loose liner of paper or polythene can be used and after transporting the commodity the liner may be taken out and the bag re-used. One of the advantages which a jute bag has over a paper bag is the fact that it has a good second-hand value and in most countries of the world there is a considerable trade in second-hand bags.

Jute is used in woven carpets as weft, warp, or pile, in tufted carpets as the backing material, in linoleum as backing, and in carpet underlays and felts. A general indication of the amounts of jute used in different floor-coverings is given below

Woven carpet (wool pile)	1·2 lb/yd^2
Woven carpet (jute pile)	2·0 lb/yd^2
Tufted carpet	0·9 lb/yd^2
Linoleum	0·6 lb/yd^2

One of the outstanding developments in the carpet industry in recent years has been the rapid growth of the tufted carpet section and now large quanties of jute are sold for the backing fabric of these carpets.

Jute is also used in smaller quanties in a host of other applications. Small domestic ropes, parcelling twines, horticultural twines are examples of its use as cordage. Roofing felt and damp courses often have a base-cloth of jute; in the upholstery trade jute is used for covering the underside of chairs and as webbing for supporting chair seats; tailors' interlinings are often made from fine jute cloth; jute yarns are used in the electrical and cable-making industries as packing for power cables or telephone and telegraph cables; jute may be used for filter cloths, boot and shoe linings, and tarpaulins. It has even had some vogue as a dress fabric.

CHAPTER ONE

The Anatomy, Cultivation, and Marketing of Jute

JUTE is obtained from the stems of two plants grown mainly in the Indian sub-continent. All fibres which are extracted from the stems of plants are classified as bast fibres, others in this category being flax, hemp, kenaf, and ramie. The botanical names of the plants from which jute is obtained are *Corchorus capsularis* and *Corchorus olitorius*. About 40 species of *Corchorus* are known throughout the world, being found chiefly in the Tropics but *C. capsularis* and *C. olitorius* are the only ones which are cultivated for their fibre. In the wild state both plants are small and shrub-like but when they are cultivated they can grow up to a height of 15 ft. Both are herbaceous annuals, i.e. they grow from seed to maturity in one year and in doing so produce seeds for the following year's crop. Jute is grown in the rainy season in temperatures of 70–100° F with relative humidities of 65–95 per cent and requires a total rainfall of about 10 in. during the months of March, April, and May.

In general appearance *C. capsularis* and *C. olitorius* are similar, having long straight stems about 1·5 in. in circumference, unbranched except at the top. The main difference between the two species is in their fruits: *C. capsularis* has a rough wrinkled spherical seed-box about 0·3 in. in diameter and *C. olitorius* has an elongated pod like a miniature cucumber about 2 in. long. Besides the shape of their seed-boxes there are other differences: *C. capsularis* tends to be shorter than *C. olitorius*, rarely exceeding a height of 12 ft compared with 15 ft for *C. olitorius*; *C. capsularis* is grown on lower-lying ground than *C. olitorius*; *C. capsularis* yields the 'white' jute of commerce and *C. olitorius* the 'Tossa' and 'Daisee'. Tossa is grown on the higher ground because the crop withstands floods later than white and so does not need to be cut at the normal flood-threat time. Although Tossa has a higher yield per acre and commands a better price, some 60 per cent of the total jute crop is of white jute.

THE ANATOMY OF THE JUTE STEM

The jute fibres lie within the stem of the plant just beneath the bark and surrounded by soft tissue. Figure 1.1 shows diagrammatically what would be seen if a V-shaped wedge were cut out of a jute stem.

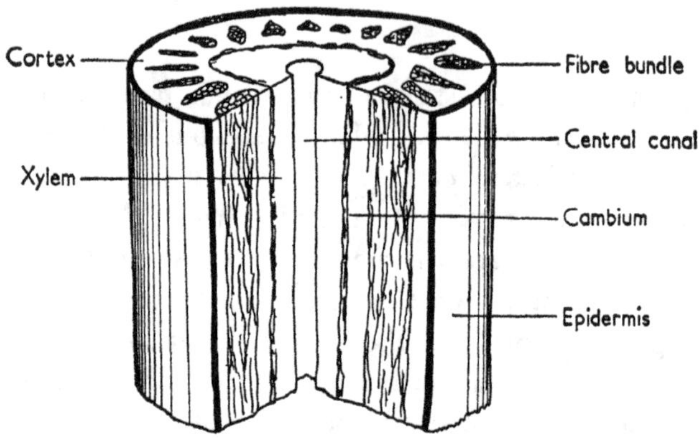

Figure 1.1. Anatomy of the jute stem

On the outside of the stem is the epidermis which in young plants is green and soft but becomes harder, particularly at the root end, as the plant matures. Immediately beneath the epidermis lies the cortex, imbedded in which are the fibre bundles. Continuing to move inwards to the axis of the stem the cambium is found, a continuous layer about five cells thick running completely round the stem. On the inside of the cambium lies the xylem which, as the plant matures, becomes more and more woody and finally, running down the centre of the stem is a canal which in mature *C. olitorius* stems is usually hollow but in *C. capsularis* stems still contains a soft pith.

The cambium plays an extremely important part in the life of the plant and is particularly interesting because it is from it that the fibre bundles develop.

In the young plant the stem is composed of a ring of unconnected bundles of cells, surrounded by the soft tissue of the cortex and encircling the pith, the whole being enclosed by the waxy epidermis which protects the young plant. The inner part of the cell bundles contain the elements of the woody xylem while the outer part will form the 'bast'; separating these two sections are the rudimentary cells of the cambium. As the plant grows, the cambium cells multiply and divide until they join up with their neighbours in the adjacent bundles to form a complete ring round the plant. On the inside of cambium the cells enlarge and bundles of them become progressively more lignified and form the 'wood' of the stem. On the outside of the cambium interesting and important developments occur. Groups of cells known as medullary rays spread out from the cambium and between them certain cells begin to change by thickening their walls—these are the first fibres. Growth continues and more and more cells develop into fibres until the easily recognizable fibre bundles are formed. The bundles are roughly triangular in shape with their base towards the cambium and their apex towards the epidermis. In the bundle the oldest fibres are at the apex and the most recently formed at the base. Thus the oldest fibres are continually being pushed outwards by the newly formed ones.

These cells in the fibre bundles are, as it were, the building units of the fibres and are called 'ultimates' and are on average about 18 microns in diameter and 2·5 mm long (1 micron = 0·001 mm). The ultimates are cemented together to form the 'fibres' of commercial usage which run along the stem of the plant, branching and dividing, only to unite with their neighbours then divide again, making up a mesh of fibre networks lying in layers around the cambium. The outermost layers are more open than the innermost ones because of the outward growth from the cambium, and as the stem circumference grows the first-formed networks become stretched and open.

THE CULTIVATION OF JUTE

The jute crop is grown on small plots of land and in many districts half the growers have only about 800 lb of fibre to sell at the end of the season. With the normal outrun of fibre being 1,100–1,300 lb per acre, this means that many of the plots are only about three-quarters of an acre. Since it takes around 80 man-days (1 man-day = 1 adult working for 7 hours) to plough, sow, weed, cut, and extract the fibre from 1 acre of ground it follows that about 150 man-days are needed to produce 1 ton of fibre. Some idea of the large labour force required can be obtained when one remembers that about 2,000,000 tons of jute are grown each year.

Low-lying, slightly acidic, alluvial soils in river complexes are particularly suited to jute growing, especially when these soils are revitalized by flooding each year and a deposit of silt is left on them when the flood-waters recede, but the fibre can be grown on lighter sandy soils provided large quantities of manure are fed into the ground. The characteristic feature of the main jute-growing areas of India and Pakistan is the low-lying nature of the terrain, any slopes are gradual and the river banks have an extremely small gradient. Dacca, in the centre of an important jute-growing region, is less than 50 ft above sea-level although it is 100 miles inland. These low lands, as would be expected, flood very easily when the heavy monsoon rains coincide with the melting of the Himalayan snows about the middle of June or July and even those parts which are not actually flooded may be under a few inches of surface water at times. The lower levels are inundated each year by the overflowing rivers which meander over the whole area and at harvest time parts of the crop may be under several feet of water. Apart from the beneficial effects of this large supply of water from the botanical and agricultural points of view, the widespread river systems provide a very useful means of transporting the fibre as road and rail communications in the country districts are not good.

The time when both types of jute cease strong growth and enter upon their reproductive phase of life by flowering and then forming seed-pods is influenced by the hours of daylight in each day. When the length of the day reduces to about 12 hr at the end of August and the beginning of September the plants flower soon afterwards no matter when they have been sown. *C. olitorius* is more sensitive than *C. capsularis* in this respect and since growth and the yield of fibre depend critically upon the time of flowering the former variety is always sown later than the latter. Most of the more commonly met *C. capsularis* is sown in February, March, or April, whereas the *C. olitorius* type is sown in April and May. Apart from these differences the two species are cultivated in similar ways.

The land is ploughed to a depth of a foot and the soil worked down to a fine tilth by successive harrowings or 'ladderings'. Laddering consists of drawing a rough bamboo ladder or a log of wood about 7 ft long across the plot with the worker standing on it to apply pressure. This breaks up the lumps of earth, levels off the soil, and removes weeds. Since jute seeds are very small (about the size of turnip seeds) they need a fine seed-bed. As jute is a strongly growing plant it requires plenty of nourishment from the soil. Where flooding occurs the fresh silt brought down each season is a ready supply of fresh nutritional material but where flooding does not occur the land must be manured.

Sowing is usually done by the broadcast method at the rate of 10 lb of seed per acre for *C. capsularis* and 6 lb per acre for *C. olitorius.* The sower walks across the field scattering the seeds to either side, then when the ground has been covered in one direction he repeats the process by walking at right angles to his original line; in this way a uniform distribution of seeds can be achieved. A light covering of earth is then drawn over the seeds until they are 1–1·5 in. below the surface, and the surface is consolidated by laddering. Line sowing, which gives a better yield of fibre is being encouraged by the various jute-growing authorities by means of field demonstrations, etc., but at the moment most seed is sown broadcast.

Within 2 or 3 days the seeds germinate and about a million plants per acre are formed. This high seed-rate is necessary because the individual seedlings are very delicate and this large number makes it easier for the plants to burst through the firm crust of earth which forms when rain follows soon after sowing. The plentiful supply of plants ensures that some will survive if periods of drought occur before the monsoon starts at the beginning of June. Weeding and thinning are carried out manually, usually in two stages when the plants are 3–6 in. tall, until a final count of around 150,000 plants are left, spaced 4–6 in. apart. Weeding is by far the most laborious part of jute growing, accounting for 30–40 per cent of all the labour involved. Depending on the district, the plants are ready for harvesting from the middle of June to the end of September.

The optimum time for harvest is just after the plant has flowered and before the fruits form since at this stage the plant has reached full height, the bark is easily retted, and the fibres are at their best. If the crop is cut early, perhaps because of heavy rains and flooding early in the season, then the yield is low, the fibre short and pale in colour; late harvesting, when the fruits are well set, gives a higher yield but the quality of the fibre deteriorates.

The plants are cut off close to the ground with a sickle and in the plots which are flooded the workers must dive beneath the water to do this. Where the water is only 2–3 ft deep the plants may be simply pulled up by the roots and then the roots cut off when the stems are on the banks. On the higher ground the stems are stacked for a few days to let the leaves fall and then they are bundled ready for the next stage in fibre extraction. Jute harvested from low ground has its stems bundled immediately after cutting.

As jute is an annual, some of the plants must be left to produce seed for the next year's crop; depending on the district some 3–5 per cent of the land is used for this purpose.

FIBRE EXTRACTION

In the living plant the fibre bundles lie beneath the bark, surrounded by gummy materials; these encircling soft tissues must be softened, dissolved, and washed away so that the fibre can be obtained from the stem. This is done by steeping the stems in water and is known as 'retting'. The bundles of stalks are laid in ponds, ditches, or slow-moving streams, weighted down with stones, leaves, or clods of earth, and left for 5–15 days. A plentiful supply of water for retting is another of the reasons why jute can only be grown on a large scale in certain regions of the world (approximately 2,800 gal are needed to pond-ret 1 ton of green stalks which will yield some 112 lb of fibre). The optimum water temperature for retting is 80° F. Retting is caused by micro-organisms which soften the tissues and gums, starting at the cambium and extending outwards so that the outer cells of the cortex are the last to disintegrate. Retting is better if the stems are uniform in thickness since large differences in diameter mean that the thin stems will be retted before the thicker ones and by removing the stems at an average time poor quality arises from the thin stems being over-retted and the thick stems under-retted. Similarly at the root end of the stem the bark is stronger and more resistant to the micro-bacterial attack than the middle of the stem which, in turn, is more resistant than the top end. The type of water which is used for retting has an influence on the value of the fibre, for instance stagnant pools where the same water is used over and over again become loaded with iron salts and the fibre is discoloured to a metallic grey shade. The best place for retting is in slow-running streams which are as free from pollution as possible. Retting, therefore, is a critical stage in the production of jute where good cultivation can be completely undone by carelessness or inattention.

When the daily examination of the stems shows that the bark can be removed easily from the rest of the stem the fibre is taken from the water as soon as possible. This stage is called 'stripping'. A bunch of stems is held in one hand and the root end tapped lightly with a mallet, this action frees the fibres at the foot of the stalk. The labourer then grasps the fibres and by jerking and lashing the stems about in the water, loosens the rest of the fibres, picks off odd pieces of bark, washes the fibre, and squeezes the excess water out. The fibre is then collected and laid out on bamboo racks to dry for 2–3 days. In some districts of East Pakistan each stem is stripped singly but although this method produces a better quality of fibre it is slow and laborious.

THE MARKETING OF JUTE

The movement of jute from the growers to the home mills or the exporters is one of collection, assembly, storage, and transportation at several different stages, each becoming a larger and more important link in the chain. The first link is the bi-weekly village market or *hat*. As the crop becomes ready in late June or early July, itinerant dealers travel round the homes of the growers buying their jute and then taking it to the *hats* where they and some of the growers who bring their own jute to the market sell to merchants. The jute at the *hat* is sold in an unassorted fashion, the only distinction being between white and Tossa jutes. The fibre is transported by country boat, pack animal, or cart to the larger secondary centres where jute buying and selling goes on daily during the season. Throughout East Pakistan there are about 250 of these secondary markets. There the fibre is graded into Tops, Middles, B-, C-, and X-Bottoms by the *kutcha* baler. A kutcha baler is one who grades the raw jute and packs it into kutcha bales weighing about 250 lb for use in the home trade. At some of the secondary markets there are *pucca* balers too but pucca baling is more commonly carried on at one of the larger terminal markets. A pucca baler grades the fibre for export, cuts off the hard root end, and presses the jute into pucca bales weighing 400 lb and measuring 49 in. × 18 in. × 20 in. This is done to save valuable cargo space for the material which is to be shipped overseas. In East Pakistan the main terminal markets are Narayanganj and Khulna and the jute is shipped through Chittagong and Calna. The home mills buy their jute either from the primary or secondary centres, the latter being the chief source.

JUTE QUALITY AND GRADING

As yet there are no objective tests made commercially on jute to assess its quality, although many experiments have been made both in India and the United Kingdom to try to relate measurable fibre characteristics such as fineness, strength, length, etc., to the properties of the yarn spun from each grade. While laboratory tests are encouraging, no commercial grading is done in this way. Quality standards have developed through the normal channels of commercial usage and though there are no rigid rules laid down by which to differentiate the various qualities there is good agreement in India, Pakistan, and overseas between experienced assessors as to fibre value. Those fibres which may be spun into fine yarns are considered to be the most valuable and those which can only be spun into coarse sacking yarns of least value.

The factors which are taken into account during grading are colour, length, fineness of fibre, lustre, strength, cleanness, freedom from defects, and the amount of root end which will have to be cut off. A strong fibre with good length, even colour, high lustre, no defects, and little root is considered good quality. Needless to say, each area produces different grades of fibre according to the prevailing soil, climatic and cultural conditions, and, particularly, the quality of the retting water available. In those parts where clean water is freely available the fibre is invariably superior to those that have only dirty muddy water. Besides retting, however, the choice of seed and the time of harvest also play their part in determining quality.

The much commoner *C. capsularis* or white jute varies in colour from pale cream to grey or yellowish-tan, the best grades having a high lustre. Tossa jute (*C. olitorius*) has a russet tinge varying from golden brown to reddish brown. Daisee jute (*C. olitorius* from the Calcutta region) is grey to black from the presence of iron salts in the retting water.

Some of the defects which may be seen in jute are as follows:

(1) Runners. Long strips of bark adhering to the stems for much of their lengths, caused by inadequate retting.
(2) Rootiness. Tough, hard, stiff pieces of bark sticking to the lower end of the fibre, caused by flood water toughening the epidermis, making it more resistant to retting.

(3) Croppy. Gummy harsh top ends to the fibre, often resulting from incomplete immersion during retting or harvesting at the wrong time.
(4) Specky. Small, black pieces of bark sticking to the fibres, due to pests or branching in the stem, both of which lead to the formation of harder bark which is difficult to ret.
(5) Dazed. Dull, weak fibre, usually limp and lifeless, caused by over-retting (or packing in damp bales).

Grading is done at two stages—one for the home trade and one for the export trade. The growers have little knowledge of jute grading and it is not until the fibre comes into the hands of the larger

TABLE 1.1 TYPES OF JUTE AND THEIR GROWING AREAS

Type	*District*	*Characteristics*
White jute (*C. capsularis*)		
Jat	Mymensingh, Dacca, Tippera	Best quality, strong, clean, lustrous, good length
District	Bogra, Pabna, Faridpur, Khulna	Medium quality, harder than Jat
Northern	Dinajpur, Pabna, Jalpaiguri	Soft fibre, medium to low quality with loose stick and speck
Western	Purnea	Soft, rather weak, discoloured
Assam	Goalpara, Nowgong, Sylhet	Variable quality
Orissa	Cuttack	Generally poor
Jungli	Murshidabad, Malda	Soft heavily rooted, poor quality
Tossa jute (*C. olitorius*)		
Jat	As for white Jat	Golden brown, strong, lustrous, clean, pliable, good length
District	Faridpur, Khulna, Nadia, Jessore	Medium quality
Northern	Murshidabad, Pabna, Bogra	Soft, lighter in colour with much loose stick and speck
Daisee jute (*C. olitorius*)		
Jat	Howrah, Hooghly, Burdwan	Long, lustrous, soft, grey
District	24 Parganas, Jessore, Khulna	Dark in colour, medium strength

merchants that it is assorted into different classes. The preliminary grading is done by kutcha balers.

The current tendency is for the growers to produce as much Tossa as possible. Some years ago the ratio of white to Tossa grown was of the order 3 : 1, but in 1962 the ratio had changed to approaching 2 : 1.

Table 1.1 gives a brief note on the characteristics of each type, and their growing areas are shown in Figure 1.2.

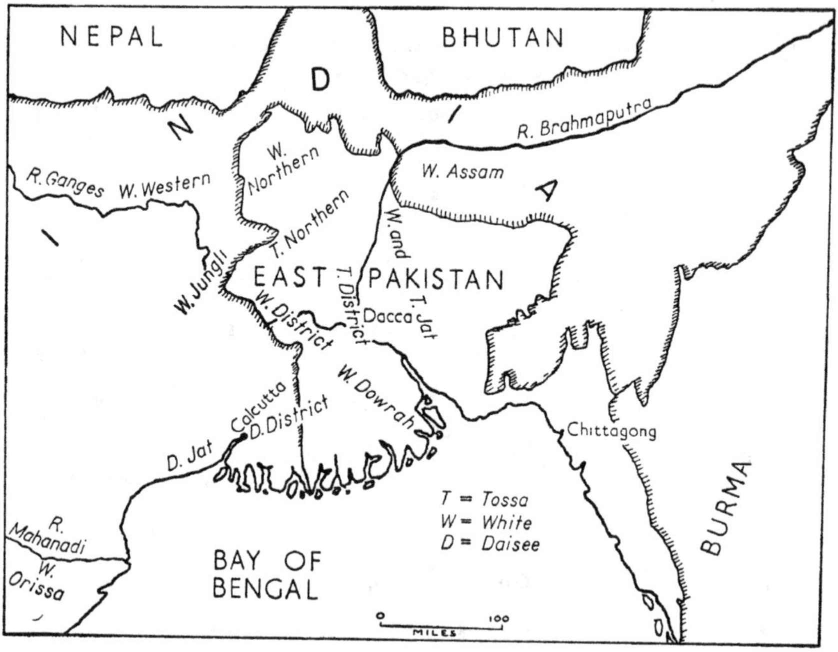

Figure 1.2. Jute growing districts of India and Pakistan

The bulk of Pakistan jute is broadly classified into three groups, Jat, District, and Northern, in descending order of merit. The kutcha baler carries out a further division of each type depending on the relative value of the fibre he has received. The criteria by which he judges the fibre are shown in Table 1.2. Nowadays, the top two gradings, i.e. Jat and District Tops, have largely vanished from the raw jute market.

Indian kutcha gradings follow the same pattern but the allowances for cuttings are about 5–10 per cent more in the top two grades and B- and C-Bottom are omitted.

TABLE 1.2. KUTCHA BALERS' GRADINGS OF RAW JUTE (1961)

Grade	*Characteristics*	*Root-end Cuttings not to exceed (per cent)*	
		White	*Tossa*
Tops	Very strong fibre, excellent colour and lustre, free from all defects	15	10
Middles	Strong, sound fibre, average colour for the district, free from speck, runners, and harsh crop end	25	15
Bottoms	Sound fibre, medium strength, free from hard-centred jute	30	20
B-Bottoms	Sound fibre, medium strength, not suitable for higher grades	30	20
C-Bottoms	Medium strength fibre, any colour, free from runners and croppiness	35	25
X-Bottoms (Cross Bottoms)	Weak, harsh jute, free from tangled jute and stick	40	30
Habijabi	Tangled, ravelled jute of any sort, free from dust and cuttings		

The other classification of raw jute is for the export trade and is done by the pucca baler. Each baler has his own house-signs or marks by which his jute may be known. Virtually all jute that is packed in pucca bales for export has the hard dark-coloured root end cut off to save the expense of root-cutting in Europe. These root ends are sold separately as cuttings.

White jute is assorted into three main classes: Crack (or Dundee), Mill, and Export (or Grade). The top class is sub-divided into Firsts, Lightnings, and Hearts; the Mill Class into Reds, Firsts, Lightnings, and Hearts; and the Export class into Firsts, Lightnings, and Hearts. Tossa jute is assorted into four classes: Dacca Tossa, Crack, Grade (or Dundee), and Outport (or Continental). Each class is sub-divided into 2/3s and 4s with 5s and 6s in the Dacca Tossa class only. Little Daisee jute appears on the export market, what there is being graded into Crack 2/3s and 4s and Grade 2/3 and 4s.

Naturally, between certain of the class divisions there is an overlap, but Table 1.3 shows the relative gradings in general terms. It must be emphasized, however, that as conditions vary from year to year the relative value of the various marks changes.

TABLE 1.3. GRADINGS OF RAW JUTE EXPORTED FROM PAKISTAN

White jute			*Tossa jute*			
Crack	*Mill*	*Export*	*Dacca Tossa*	*Crack*	*Grade*	*Outport*
Hearts			4			
	Firsts		5	4	2/3	
	Lightnings	Firsts	6		4	
	Hearts	Lightnings				2/3
		Hearts				4

The link between the kutcha assortment and that of the pucca baler is roughly as in Table 1.4.

TABLE 1.4. APPROXIMATE RELATION BETWEEN KUTCHA AND PUCCA GRADES

Kutcha grades	*Pucca grades*
White jute	
District middle and Northern top	Crack Hearts
Jat bottom and Northern middle	Mill Firsts
District bottom	Mill Lightnings
Jat X-Bottom and Northern bottom	Mill Hearts
All jute not in above classes	Grade Hearts
Tossa jute	
Jat middle	Dacca Tossa 4
Jat bottom	Dacca Tossa 5
Jat X-bottom	Dacca Tossa 6
Northern top	Grade Tossa 2/3
District middle	Crack Tossa 4
District bottom and Northern middle	Grade Tossa 4
Northern bottom	Outport Tossa 2/3
All jute not in above classes	Outport Tossa 4

RAW JUTE MOISTURE

In the normal course of events the fibre is saturated with water when it leaves the retting pits and must be dried off before it can be sold. In its passage from up-country to the baling centres it may become wet again and need to be dried before baling. There is no standard moisture content for baled jute but claims for damage due to excessive moisture can be taken to arbitration. In Pakistan legislation now exists which prohibits the sale or purchase of damp jute.

The most serious effect of excessive quantities of moisture in baled jute is 'heart damage'. When a heart-damaged bale is opened it is found that fibre in the centre of the bale is brittle and powdery. This damage results from micro-biological activity and only occurs under certain conditions of temperature and moisture. This action is unlikely to begin if the moisture content of the bale is below 19–20 per cent no matter how high the ambient temperature becomes, but at moisture contents in excess of this level there is a danger that damage may occur. It will be appreciated that in passage *en route* to Europe through the Red Sea the temperatures in a ship's hold become very high. In bales susceptible to heart damage these temperatures can stimulate bacterial growth. In spite of recourse to arbitration, the loss to the buyer resulting from heart damage is a serious one.

TABLE 1.5. MOISTURE CONTENT OF PUCCA BALES OPENED IN U.K.

Annual figures		*Monthly figures*	
Year	*Average moisture content (per cent)*	*Month of opening*	*Average for* 1954–59 *(per cent)*
1954	13·3	January	14·0
1955	13·5	February	14·0
1956	13·9	March	13·9
1957	13·8	April	13·6
1958	13·4	May	13·3
1959	13·2	June	13·1
		July	13·0
		August	13·1
		September	13·2
		October	13·4
		November	13·8
		December	14·0

Table 1.5 shows the results of tests carried out on raw jute imported into the United Kingdom over the years 1954–59. The points to be noted are that bales opened between November and March have a slightly higher moisture content than those opened during the rest of the year. This represents the arrival of a new crop. Taken on

average there is very little variation from year to year, but this is not to say that individual bales are equally uniform—moisture contents as high as 22 per cent and as low as 10 per cent are not unknown.

ALTERNATIVES TO JUTE

Many countries have tried to develop plants which could be used as a substitute for jute but at the present time only three plants can be considered as commercially successful in this respect. All belong to the same botanical family as jute and, in many respects, are very similar to jute itself.

Hibiscus cannibinus is grown widely in the Tropics and sub-Tropics where it is known by many names, e.g. Bimli jute, mestha, Deccan hemp, stockroos, or kenaf. It can stand lower temperatures than jute and is grown as far north as the Caspian Sea and as far south as the Transvaal in the Republic of South Africa. The most important growing region, however, is India, mainly Hyderabad and Madras in the south and Bihar in the north. In general, cultivation follows the same pattern as jute but there have been more attempts to harvest the crop mechanically and to extract the fibre by machine.

Hibiscus sabdariffa is being grown on a fairly large scale in Thailand and other parts of the world. The fibre is properly called 'roselle' but in Thailand it is known as kenaf and when the same fibre reaches Europe it becomes 'Siamese jute'—a clear demonstration of the confusion which exists in the nomenclature of the jute alternatives. Many breeding trials have been made with *H. sabdariffa* with the object of producing tall, disease-resistant strains. The properties of roselle are similar to jute and the fibre may be spun either alone or in a mixture with jute without any modifications to the machinery.

Urena lobata, Congo jute or aramina, is grown chiefly in the Congo and now supplies that country with much of its fibre for sack requirements. Cultivation is easy and the yield is high, retting being carried out in the same way as jute retting. At present little of this fibre comes on to the world market as most of it is used internally.

CHAPTER TWO

The Structure and Properties of Jute

WHEN jute is extracted from the bast of its parent plant it is in the form of a long mesh of interconnecting fibres, in some places compacted into a flat ribbon and in others opening out into a network. This is the smallest unit of the commercial raw jute trade and is known as the reed, i.e. the aggregation of fibres coming from the stem of one plant. The reeds may be 3 to 14 ft long, depending on the grade, and they show quite clearly the taper of the stem from root to crop. Generally, long reeds have thicker root ends than short ones. Reeds which are thick all along their length tend to give coarse fibres and thin reeds tend to give fine fibres.

During manufacture the reeds are opened out and split into their component fibres; these are the fibre entities as far as the manufacturer is concerned. The weight per unit length of individual fibres varies from 0·7 to 5·5 tex but the average is between 1·9 and 2·2 tex. There is no clearly defined average fibre length and any sample of jute fibres contains large numbers of short fibres and a few long ones. Figure 2.1 shows the type of fibre length diagram obtained from jute; this pattern, combined with the relatively large diameter of the fibres, confines the use of jute to the heavier counts of yarn in comparison with wool, cotton, or flax, for if fine yarns are to be spun then fine fibres are a necessity.

The spinner's fibre is, in turn, composed of a number of smaller cells—the ultimates. There are usually between 6 and 20 ultimates in each cross-section of a fibre and in diameter they range from 6 to 20 microns and in length from 0·7 to 6·0 mm with an average of 2·5 mm. The cell walls are thick and in the centre of the cell is a hollow lumen which, in life, is filled with protoplasm. The lumen is irregular in cross-section sometimes becoming broad, making the cell walls thin at that point. Plate I shows a transverse section and a longitudinal view of a jute fibre. The characteristic shape of the ultimates can be seen.

Jute, like most of the other textile fibres, is hygroscopic, i.e. it takes in or gives out moisture to its surrounding atmosphere. This it does

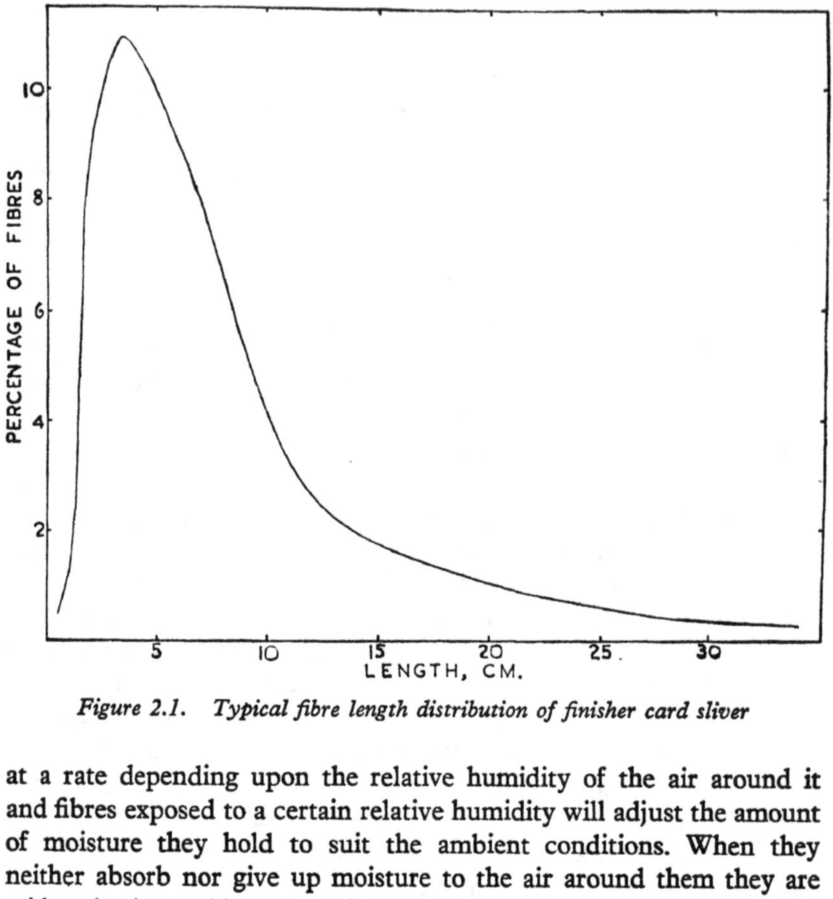

Figure 2.1. Typical fibre length distribution of finisher card sliver

at a rate depending upon the relative humidity of the air around it and fibres exposed to a certain relative humidity will adjust the amount of moisture they hold to suit the ambient conditions. When they neither absorb nor give up moisture to the air around them they are said to be in equilibrium with that particular atmosphere. Thus jute freely conditioned in certain ambient conditions contains a specific quantity of moisture. The amount of moisture held by jute can be expressed in two ways, by moisture content or moisture regain:

$$\text{Moisture content (\%)} = \frac{\text{Weight of moisture present} \times 100}{\text{Total weight of sample}}$$

$$\text{Moisture regain (\%)} = \frac{\text{Weight of moisture present} \times 100}{\text{Weight of bone-dry fibre}}$$

For reasons which will appear later, moisture regain is to be preferred.

If a sample of jute is split into two halves, one of which is dried in an oven and the other soaked in water, and the two allowed to condition

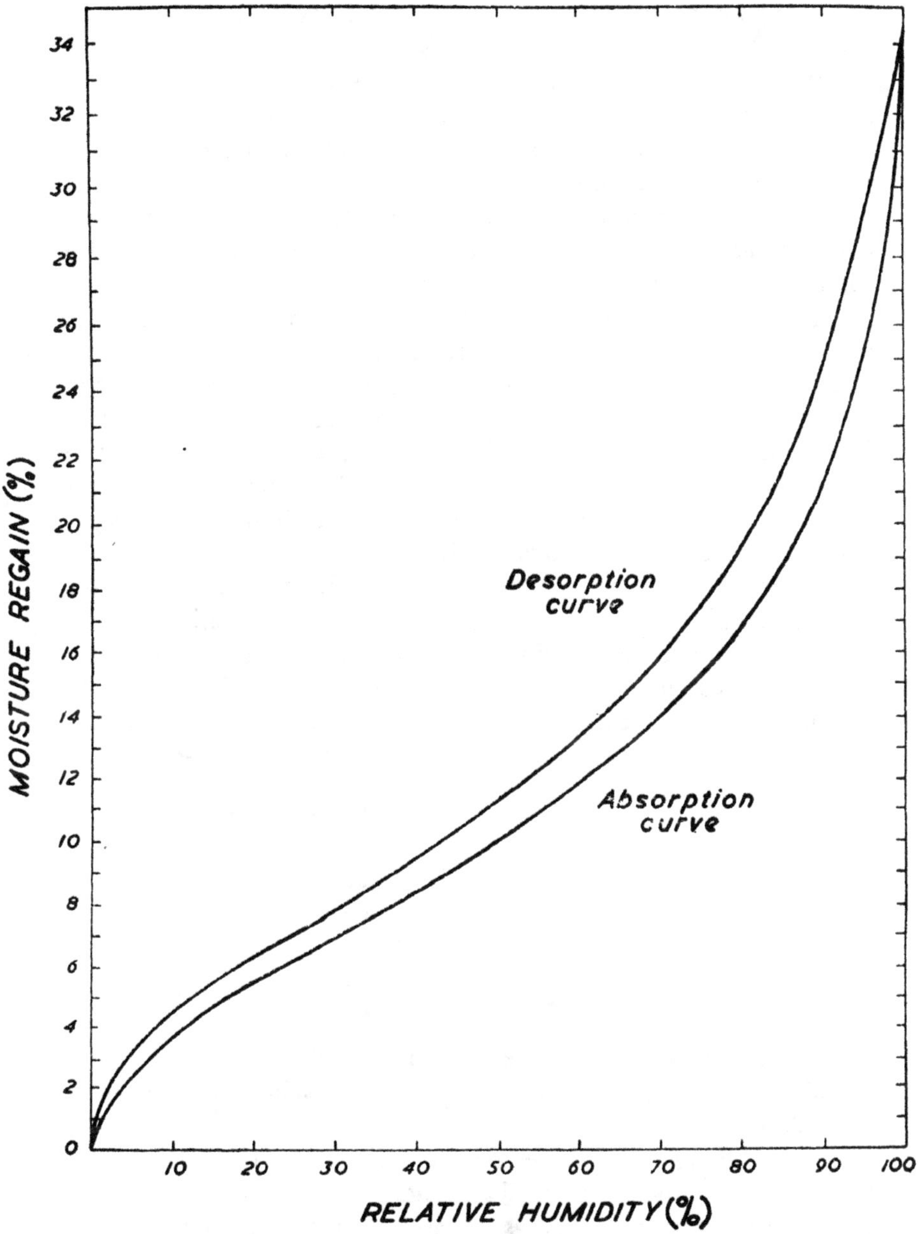

Figure 2.2. Moisture regain of jute at various humidities

in a certain atmosphere, they will not ultimately contain exactly the same amount of moisture. This is a result of a hysteresis effect. Figure 2.2 shows the moisture regains of jute conditioned at various humidities, with the hysteresis loop clearly shown. If jute approaches its equilibrium regain from the 'wet' side it is said to be on the desorption part of the curve and if it approaches it from the 'dry' side it is on the absorption part of the curve. At the standard relative humidity of 65 per cent at 20° C the two regains are about 2 per cent apart. In normal conditions of jute spinning the fibre always approaches equilibrium from the 'wet' side.

It may be mentioned in passing that moisture has another effect on the fibre which is used as a basis for measuring the moisture regain of the material. As the moisture regain increases, the electrical resistance of the fibres becomes less and their dielectric constant increases. These phenomena are utilized in two types of electronic moisture meters which measure the resistance or the dielectric constant of the fibres. These measurements can then be converted into moisture regain figures by consulting a calibration chart.

The tensile strength of a textile material is not an absolute figure but depends upon the well known influence of test length, method of loading adopted, machine capacity, etc., which the reader will find discussed in text books on testing. The effect of test length on the tenacity of jute is illustrated by the following figures taken from data by Mukherjee, Sen, and Wood (*J. text. Inst.* **39**, P243 (1948)).

Test length (in.)	*Tenacity (g/tex)*
0·5	60
1·0	49
1·5	46
2·0	43
2·5	40

Table 2.1 shows comparative tenacities for a selection of bast and leaf fibres (6 mm test length, constant-rate-of-loading machine, time-to-break 10 sec).

Chemically, jute has three main constituents: (1) Cellulose; (2) Hemicellulose; (3) Lignin. Small amounts of nitrogenous and inorganic material are present as well as variable amounts of water. So that differences in the percentage composition may not result simply from different levels of moisture in the fibre it is customary to express the

TABLE 2.1. FIBRE TENACITIES AND EXTENSIONS AT BREAK

Fibre	*Tenacity (g/tex)*	*Extension at break (per cent)*
Jute	70	2·0
Hemp	84	4·2
Ramie	80	3·9
Sisal	52	5·0
Manila	65	7·8

analysis on an oven-dry basis. This has been done in Table 2.2. where the chemical compositions of jute and some other fibres are compared.

TABLE 2.2. CHEMICAL COMPOSITION (PER CENT) OF JUTE AND OTHER FIBRES

Fibre	*Cellulose*	*Hemi-cellulose*	*Lignin*	*Pectin*	*Water-solubles*	*Fats and waxes*
Jute	65·2	22·2	10·8	—	1·5	0·3
Hemp	77·5	10·1	6·8	2·9	1·8	0·9
Ramie	76·6	8·0	5·6	3·8	5·6	0·4
Sisal	71·5	18·1	5·9	2·3	1·7	0·5
Flax	71·3	18·5	2·2	2·0	4·3	1·7
Cotton	91·5	5·8 (with pectins)	—	—	1·1	0·7

The cellulose, hemicellulose, and lignin all exist in the form of long-chain molecules, as indeed do all the principal chemical compounds in textile fibres. Just as the individual fibres in a yarn are long and thin and hold together by a mixture of entanglement and inter-fibre friction so the long-chain molecules in the fibre are long and thin and hold together by a mixture of entanglement and chemical forces. The long-chain molecules can be likened to a string of beads, each bead being the characteristic building unit of the molecule.

Cellulose is the only 'pure' substance to be found in jute and the cellulose extracted from the jute fibre is identical with that found in all other cellulosic fibres. The building unit of the long-chain cellulose molecule is the simple sugar, glucose, which has been made by the plant from the elements carbon, hydrogen, and oxygen.

The hemicellulose molecule is made up of smaller units, just as pure cellulose, but in this case the 'beads' are different types of sugars and the chains are very much shorter. Another difference between the two types of cellulose lies in the shape of the long-chain molecules; the cellulose chain has many identical glucose units strung head-to-tail but the hemicellulose molecule has short side-chains sticking out at intervals along its length. These side-chains are acidic in nature and it is they which give jute its slightly acid reaction and its affinity for basic dyestuffs.

Lignin differs from the other two main components of jute in not being made up from sugar units. Lignin is an important constituent of wood and though its chemical structure has been under examination for more than 100 years its exact nature has not yet been established. Jute is the most highly lignified fibre of commercial importance, a feature which determines many of its characteristic properties. For instance, the strength of the fibre is higher than would be expected from an examination of the molecular structure of the fibre, and it is thought that this is due to the lignin molecules forming linkages which help to give the fibre additional strength. These same linkages, however, reduce the flexibility and extension of the fibre. Lignin, too, is thought to be responsible for the wide colour range of the fibre—far wider than that of any other textile fibre. The yellowing of jute on exposure to sunlight is due to the lignin, while the fibre's good resistance to bacterial degradation is another example of this compound's important role in determining some of the properties of jute.

Jute, in common with many of the other textile fibres, may be degraded by sunlight, heat, mildew, acids, and alkalis, but by the choice of suitable reagents the fibre's resistance to these damaging influences may be improved and the life of the product prolonged considerably. Treatment with copper salts gives jute a good resistance to microbiological attack although this treatment is not recommended for material which will ultimately be used for packing foodstuffs. When jute is exposed to acid fumes, as it may be when used to pack some types of fertilizers, a pre-treatment with sodium benzoate will provide adequate protection to the fibre or a paper or polythene liner may be used in the sack to keep the acid fumes away from the jute. Jute will ignite and burn but its flammability may be reduced by treating it with a borax and boric acid mixture, antimony salts, or other media, and finishes are available which inhibit flaming and smouldering and will withstand immersion in sea-water for 6 months. To reduce the damage done to jute by sunlight, copper salts and certain dyestuffs, e.g. Chlorazol Brown M.S., may be used. There is, however, another phenomenon connected with sunlight—yellowing. When jute is exposed to the light it gradually assumes a yellowish tinge. This, as has already been indicated, is due to colour changes within the fibre connected with the lignin molecules. It may be made more obvious by an additional factor, viz., discoloration of the mineral oil applied to lubricate the fibre during manufacture. If jute has been used as a base cloth for polyvinyl chloride coatings, the oil will gradually migrate into the plastic and turn yellow on exposure to light. This defect is normally associated with light pastel shades of coating. The remedy is to reduce the quantity of oil which is added at spinning to 1 per cent or less, or to use a more highly refined oil which will not yellow. e.g. technical white oil.

The main properties of jute can be summarized as

Ultimates	
length	2·5 mm
diameter	18 microns
Single fibres	
length	0·2–30 in.
tex	1·9–2·2
tenacity	40–70 g/tex
extension at break	2·0 per cent
Moisture regain at 65 per cent R.H.	
absorption	12·8 per cent
desorption	14·6 per cent
Specific gravity	1·48

CHAPTER THREE

An Outline of the Process

ALTHOUGH jute had been used for many years in India for making cloth, the fibre was not known in Europe until the last years of the eighteenth century. Small quantities of jute were brought to England by the East India Company in 1796 and were sent to Abingdon in Berkshire, then an important centre of the twine trade. There it was spun by hand and used in a small way in the local manufactures. At this time Scotland, and in particular Forfarshire, had extensive trade in flax fabrics, and in 1823 a bale or two of jute was bought by a Dundee merchant. However, the local spinners were not impressed and it was largely due to the foresight and tenacity of one or two merchants that small parcels were occasionally brought to the town and finally in 1832 or 1833 a spinner succeeded in making an acceptable yarn. For the first year or two little pure jute yarn was spun, mixing with flax being considered essential but after it was discovered that the fibre could be spun more easily if water and oil were added, then yarns made wholly from jute became more and more popular. Dundee, being a whaling port, had plentiful supplies of whale oil and this was the oil used in jute spinning. The trade progressed rapidly and by the middle of the century 20,000–30,000 tons a year were being used in Britain, chiefly in Dundee and district, and by 1900, 277,000 tons were being used. The first Indian mill was set up in 1855, to be followed by a rapid growth of other mills in the Calcutta region. By 1885, 7,000 looms were working.

THE SPINNING PROCESS

In general terms the types of jute yarns manufactured can be classified according to the use to which they will be put.

(1) Fine Yarns: low count yarns for making fine fabrics for tailor's inter-linings and the like. The volume of trade in these is comparatively small since they are expensive and the top grades of jute must be used to enable such yarns to be spun.

(2) Hessian qualities: medium weight yarns for weaving cloths for general packing purposes, linoleum backings, carpet backings, etc.
(3) Carpet Yarns: usually medium/heavy weight yarns of good quality either single or two-ply for the carpet industry.
(4) Sacking Yarns: medium/heavy yarns of lower grade used for the manufacture of sacks and bags.

Types (1), (2), and (4) can be divided into warp and weft qualities, the warp being superior to the weft as it must withstand the tensions of weaving while the weft acts more as a filler and undergoes little strain.

The spinning process depends upon which class of goods is being made but there are features common to all systems, viz., all jute must be softened and lubricated with oil and water so that the fibre may be processed without excessive fibre breakage and waste; the meshy nature of the reeds must be split up and the fibres separated as far as possible; the fibres must be drawn evenly into a sliver or loose untwisted strand which is then drawn out to the desired thickness of yarn; the fibres must be twisted together to give cohesion and strength to the yarn.

FINE, HESSIAN, AND CARPET YARNS

These classes of yarn are made from long jute, i.e. jute from which the root ends have been cut. The first requirement is that several different types of jute be blended together so that long runs of uniform quality can be achieved and the desirable properties of the various types of jute can be utilized and the cost of the raw material kept to a reasonable level. If the jute comes from a pucca bale it is hard and stiff after being subjected to the high pressure of the baling press and must be made more pliable before any further processing can be carried out. This is done by passing the jute through a machine called the bale-opener which has two or more heavy fluted rollers between which the jute is fed. In its passage through this machine the jute is flexed back and forth and emerges quite pliable at the other side. At this stage, however, the fibres are still rather harsh and brittle and must be softened and lubricated before they can be further processed. This is done at a machine called the spreader which consists basically of two endless chains carrying heavy pins, one chain running faster than the other. The jute is fed on to the pins of the slow chain and traverses the machine until it is gripped by the pins of the fast chain which tease and comb out the reeds. At the other end of the fast chain an emulsion of water and oil is applied, then the jute is wound up into a roll under heavy pressure. Usually the oil is a mineral oil of the light spindle variety but some of the fine yarns are still lubricated with whale oil. After this the jute is laid aside for one to two days to allow the water and oil to spread more evenly throughout the rolls of sliver.

The material at this stage is still visible in the form of reeds and the next step is to open up the reeds and separate the fibres. This is one of the functions of carding. Two cards are employed, the breaker and the finisher, each consisting of a large central cylinder covered with small sharp pins with a series of smaller pin-covered rollers set around its periphery. As the jute is fed into the machine it meets the rapidly moving cylinder pins and is combed and teased out. As it passes further round the machine it comes into contact with the pins on the smaller rollers which continue the combing, splitting, and opening action and by the time the jute has been put through both cards it is in a finely divided state showing no signs of the original fibre complexities at all. Two additional functions of the cards must be mentioned here, drafting and doubling. If jute is fed on to a pair of rollers which have a surface speed of 5 yd/min and then moved forward to meet another pair of rollers which have a linear speed of 40 yd/min then the jute will be drawn out, or drafted, and the fibres will slip past one another. The amount by which they will slip past each other, the draft, depends upon the relative surface speeds of the two rollers and in the example quoted the draft would be 8 ($40 \div 5$) and each yard of sliver going in would be drawn out to 8 times its original length. Since there has been no change in the total weight of jute it follows that if the sliver is 8 times longer than it was at the beginning then it must be 8 times thinner. These are the two important features of drafting which will be referred to again and again at all stages since they are vital to the spinning process.

The other function of the cards is to provide doublings. In all textile yarns it is desirable that the weight of the yarn should be the same, or nearly so, at all points; if some parts are very thick and others very thin then the yarn will be of low value. If one examines the sliver issuing from the spreader it will be readily seen that this desirable regularity is conspicuous by its absence, but if one places several such slivers side by side it is immediately apparent that some of the thick places coincide, purely by chance, with some of the thin ones and the resultant product is more uniform in weight along its length.

This is known as doubling and usually at each stage in the process several slivers are fed into each machine at the same time so that the thicks and thins in them will be evened out. Commonly 6 to 8 slivers are fed to the breaker card and 10 breaker card slivers to the finisher card.

After carding, the yarn is given two, three, or four passages through drawing frames. These are machines which continue the drafting and doubling begun at the cards so that by the time the material emerges from the last drawing frame it weighs about 1 lb per 100 yd. For the fine yarns four drawing passages are usual so that the slivers are drafted in easy stages and a large number of doublings can be obtained, for not only do these yarns demand the best grades of jute but the material presented to the spinning frames must be as even as possible. The last drawing passage in this case is done on a machine called the roving frame. As the sliver is now in such a tenuous state some slight degree of twist must be put into it so that it will hold together; this twist is inserted at the roving frame by inverted U-shaped flyers which rotate at about 800 r.p.m., twisting the thin sliver into a rove as they do so. The rove is wound on to a bobbin on the roving frame and is ready for the final stage of spinning.

Hessian yarns are given two or three drawing passages, the latter number being commoner. Just as the sliver at the roving frame is thin, tenuous, and weak, and must be twisted, so the sliver emerging from the final drawing frame must be strengthened to allow it to be handled. This is done by crimping the sliver, i.e. forcing small waves or crimp into the fibres to increase their grip on one another and give stability to the sliver.

At the spinning frame the material is given its final drafting down to the required weight of yarn and the fibres are twisted together to form the yarn, which is then wound up on bobbins. Twisting is done by flyers rotating at speeds of 3,500–4,000 r.p.m.

SACKING YARNS

Sacking qualities are made from poorer quality jute, the weft being composed of cuttings, mill wastes, and low-grade long jute. Because of the short nature of the raw material the spreader cannot be used, so the fibre is fed into a machine known as the softener which comprises about 70 pairs of fluted rollers. As the jute passes between these rollers it is flexed and, as the name of the machine implies, softened.

As the jute comes along the rollers the emulsion of oil and water is applied. At the exit from the softener the hard root ends of the long jute are cut off, the roots being used for weft and the remainder of the reeds for warp. The warp material is laid aside to mature for 24 hr and then is fed to a breaker card and a finisher card. The root ends, plus additional supplies of cuttings from the hessian grades, bale ropes, and other low-grade materials are softened and then matured for up to 10 days—a longer period being required because of the barky, dirty nature of the jute. The sacking weft material is given a preliminary carding in a teaser card. This machine is similar to a breaker card but with stronger, more rugged pins to cope with the hard material. The jute issues from the teaser as a fleecy tow which is then fed to the sacking weft breaker card, along with mill waste and rejections from higher grades. A finisher card follows the breaker in the usual way. Sacking warp and weft is given only two drawing passages and then is spun into yarn on large flyer spinning frames.

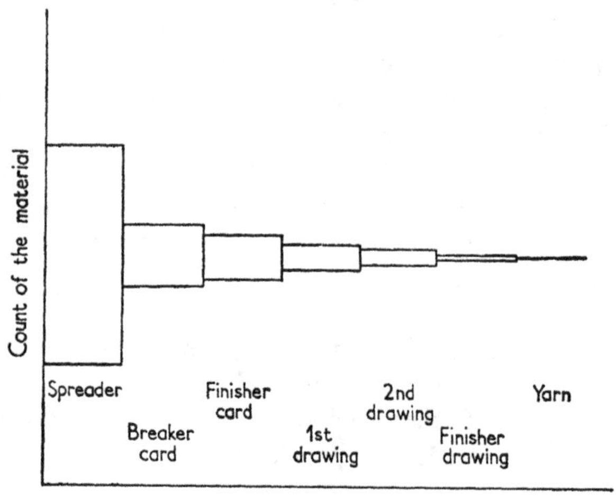

Figure 3.1. Relative counts in the jute process

Figure 3.1 shows the relative sizes of the slivers at the various stages in a hessian mill, giving some indication of the amount of drafting which must be done in reducing spreader sliver with some 137,000 fibres in its cross-section to a yarn with only about 140 fibres in its section.

COUNT SYSTEMS

The traditional units for describing the weight per unit length, or 'count', of jute slivers, roves, and yarns are as follows.

Sliver: pounds per 100 yd
Rove and yarn: pounds per spyndle
(1 spyndle (sp) = 14,400 yd)

In this volume the tex system will also be used. The count in tex being the weight in grams of 1 kilometer of material. Since jute slivers may be as heavy as 320,000 tex, the term kilotex will be used where appropriate (1 ktex = 1,000 tex). The factors for conversion from one system to the other are

$$\text{ktex} = 5 \times \text{lb}/100 \text{ yd}$$
$$\text{tex} = 34{\cdot}5 \times \text{lb}/\text{sp}$$

The range of yarns spun from jute is indicated in Table 3.1, and Figure 3.2 gives a summary of the different manufacturing systems.

FINE YARNS	MEDIUM YARNS	SACKING WARP	SACKING WEFT
Bale selection, top quality essential	Bale selection, medium grade	Bale selection, lower grades	Root cuttings, bale ropes, tangled fibre
Bale opening	Bale opening		
Spreader—water and oil applied	Spreader—water and oil applied	Softener—water and oil applied	Softener—water and oil applied
Stand at least 48 hr.	Stand 24–48 hr.	Stand 24 hr.	Stand up to 10 days
			Mix with mill waste
			Teaser card
			Mix with long jute (X-bottoms, etc.)
Breaker card	Breaker card	Breaker card	Breaker card
Finisher card	Finisher card	Finisher card	Finisher card
First drawing	First drawing	First drawing	First drawing
Intermediate drawing	Intermediate drawing (optional)		
Finisher drawing	Finisher drawing	Finisher drawing	Finisher drawing
Roving frame			
Spinning	Spinning	Spinning	Spinning

Figure 3.2. Flowsheets for jute spinning

TABLE 3.1. JUTE YARNS

	tex	*lb/sp*
Fine Yarns	120–200	3·5–6·0
Hessian Warp	240–300	7– 9
Hessian Weft	240–400	7–12
Sacking Warp	270–350	8–10
Sacking Weft	700–1400	20–40
Carpet Yarns	480–820	14–24

CHAPTER FOUR

Jute Batching Oils and Emulsions

If the jute fibre is taken from the bale and passed directly over the spinning machinery, then the yarn which is made from it is weak and irregular and the amount of waste in processing is high. In order to produce an acceptable yarn it has been customary from the earliest days of jute manufacture on an industrial scale to condition the fibre for spinning by adding oil and water to it—this operation is known as batching.

The water softens the fibre and increases its extensibility, both of which factors prevent excessive fibre breakage at the cards, make it easier for the fibre to bend round pins and rollers, and reduce waste losses. The exact nature of the part played by the oil is not fully understood but it is thought to have an important role in giving cohesion to the slivers, helping them to be drafted properly at the later stages. Originally, the oil and water were added separately but now they are, almost invariably, added simultaneously as an emulsion.

JUTE BATCHING OILS

The requirements of a good batching oil are as follows

(1) It must have no harmful effect on either the jute or the machines.
(2) The colour must be acceptable.
(3) There must be no danger of spontaneous combustion.
(4) It should not go rancid or sticky on standing.
(5) It should not have an objectionable odour.
(6) It must be cheap and in plentiful supply.

In the early days of the industry whale oil was used extensively for batching, mainly because Dundee was a whaling port at that time and there was a copious and cheap supply of this type of oil, but now mineral oil of the light spindle variety is used almost exclusively, although small amounts of whale oil are still used in spinning fine yarns.

Most jute yarns are spun with 5–6 per cent oil but for special purposes it may be necessary to reduce this to 1 per cent or less. Yarns with an oil content as low as 1 per cent are more expensive than those with the higher oil contents because to arrive at the same total weight extra fibre must be added to compensate for the reduced quantity of oil present, and raw jute is costlier than mineral oil. On the cost of materials alone these yarns must be sold at a higher price, quite apart from any additional processing costs. From this it may be deduced that it is economically desirable to work with as high an oil content as possible, but there is a technical limit to the amount which can be added. Oil contents much in excess of 6 per cent cause difficulty because the pins, conductors, and rollers of the machinery become coated with a black dirty deposit which can lower the quality of the jute passed over them.

Once the oil level has been decided the quantity of oil required is calculated from the weight of fibre which is to be put through the system. Some of this added oil, however, is lost in the fibre wastes beneath the machines. This waste is always heavily loaded with oil, particularly at the cards. The amount of oil which is lost varies somewhat but a common amount is 10 per cent of the oil that was added at batching; thus if 20 lb of oil had been added to 400 lb of jute, at the end of the process only 18 lb of oil would be left in the yarn.

Though jute spinning without oil is not a commercial proposition the amount of oil and its nature do not appear to be critical. Table 4.1 gives the results of tests in which the oil content of the yarn was changed from 0·5 per cent to 9·0 per cent and the strength of the yarn examined.

TABLE 4.1. EFFECT OF OIL CONTENT ON THE STRENGTH OF YARNS

Oil content (per cent)	*Tenacity (g/tex)*
0·0	12·2
0·5	14·0
1·0	14·6
2·5	14·4
5·0	14·5
9·0	13·9

It can be seen from the Table that as long as some oil is present the tenacity of the yarn is not altered significantly, and even if almost twice the usual amount is added the strength is not changed radically. Other tests showed that although oils of high viscosity do lower the breaking load of a yarn and make it more irregular, these effects do not become apparent until the oil is two or three times as viscous as normal batching oil.

END-USE PROBLEMS ASSOCIATED WITH BATCHING OIL

As time passes and the uses of jute become more widespread and specialized it is sometimes necessary to use lesser amounts of oil or different types of oil. Examples of such end-uses are found in the packing of food, carpet yarns, and tufted carpet backings. Some foodstuffs, such as flour, are in their final state when they are packed in jute bags and it is important that the contents do not become contaminated in any way with the batching oil since this may impart an oily taste. The best answer to this particular problem is to use a more highly refined oil which is tasteless and odourless, although reducing the normal batching oil content to 1 per cent or less will also meet the case in most circumstances. In carpets, be they woven or tufted, where a viscose or cotton pile is used it is usually necessary to reduce the oil content to 1 per cent or below to avoid the problem of soiling. Soiling of the pile yarns is caused by the batching oil 'wicking' the pile fibres and holding the dust particles in their crevasses. With the low oil content material, wicking is reduced to such a level that soiling does not occur. If a jute fabric is to be used as a base cloth for PVC coating then it sometimes happens that the oil migrates into the plastic and gradually turns yellow forming diffuse stains; these are particularly noticeable with pastel shades of PVC. This may be overcome by using 1 per cent or less of normal batching oil or changing to a technical white oil which does not turn yellow on exposure to light.

In these problems certain factors are common. The problem is associated with the batching oil, and that at the usual 5 per cent level; in all cases the difficulties may be overcome either by reducing the amount of oil present or changing to a more highly refined oil; the reason for the trouble lies in the transfer of the oil from the jute to the other material. The transfer rate depends chiefly on the following.

(1) The oil content. The higher the oil content the faster is the transfer; in fact oil migrates from jute at a speed depending on the cube of the oil content, i.e. the amount of oil migrating from two cloths with 4 and 6 per cent oil, respectively, will be in the ratio of $4^3 : 6^3$ or 64 : 216 (about 1 : 3½).

(2) Time. The quantity of oil which will pass from the jute to the neighbouring material depends on the time of contact and varies as the square root of the time; in other words, if a certain amount migrates in 1 week, twice that amount will migrate in 4 weeks, three times as much in 9 weeks, and so on.

(3) The nature of the absorbing material. Fine powders absorb more oil than coarse ones, low tex pile fibres more than high tex ones, etc.

It will have become apparent from the previous paragraphs that if these problems are to be overcome then the jute must not soil or stain the material it is in contact with. Yarns of low oil content (1 per cent or less) are therefore called 'stainless'. Table 4.2 gives a summary of the use of oils required for jute's more specialized applications.

TABLE 4.2. JUTE BATCHING OILS FOR SPECIAL PRODUCTS

End-use	*Problem*	*Oil addition*
Carpet backings for all pile yarns, except 100% jute, 100% wool, 100% nylon, wool/nylon mixtures	Soiling	Less than 1%
Upholstery	Stains on hide or PVC	Less than 1%
Packing foodstuffs	Tainting	3·5–4·5%
Packing easily tainted food, e.g. flour, sugar	Tainting	5% odourless oil
Base cloth for PVC	Yellowing	Less than 1% normal oil or 5% technical white oil

JUTE BATCHING EMULSIONS

Jute batching emulsions are of a simple nature, usually containing only the mineral oil, water, and an emulsifying agent. An emulsion is an intimate mixture of two immiscible liquids, one dispersed in droplet form inside the other. This apparently contradictory definition requires expansion. Oil and water are usually immiscible but if the oil can be split up into minute drops which are prevented from coalescing then they can be dispersed throughout the water—this is then called an emulsion. Emulsions are said to have two phases, an external phase and an internal phase. In jute batching emulsions the external phase is water and the internal phase (which is dispersed through it) is the mineral oil. Batching oil is normally a golden, amber colour and water is, of course, colourless, but when the two are mixed as an emulsion the resultant liquid is milky white. The reason for this lies with the extremely small oil droplets which scatter the light in all directions giving a white appearance, just as ground glass appears white because of all the minute pits in its surface.

A typical batching emulsion may have 30 gal of mineral oil and 80 gal of water and the problem is to split this 30 gal of oil into microscopically small drops and then disperse them throughout the water in such a way that they will stay as droplets and not reconstitute themselves into one mass of oil.

Some idea of the task may be had from the following calculation. If the water and oil were poured in a cylindrical tank 2 ft (60·96 cm) in radius the two will mix crudely but very quickly the oil will float to the top of the tank because its density is lower than that of water, and a well-defined boundary will form between the two liquids. This boundary is called the interface. The area of the interface between the oil and water in this tank will be 11,669 cm^2 (πr^2). Now if this amount of oil is to be emulsified successfully it must be transformed into droplets which are about 5 microns in diameter and it is interesting to calculate how many drops there will be.

$$\begin{aligned}
\text{Total volume of oil} &= 30 \text{ gal} \\
&= 136{,}290 \text{ cm}^3 \\
\text{Volume of one drop} &= \tfrac{4}{3}\pi r^3 \\
&= 65{\cdot}4 \times 10^{-12} \text{ cm}^3 \\
\text{Number of drops} &= \frac{\text{Volume of oil}}{\text{Volume of drop}} \\
&= 2{\cdot}1 \times 10^{15}
\end{aligned}$$

i.e. 2,100 million million drops

The original interface was 11,669 cm^2 but once the drops have been made it has been increased about 600,000 times.

The first part in the work of making an emulsion is to split up the oil into this vast number of extremely small drops. This is done by agitating the oil violently and the greater the amount of energy put into the preparation of the emulsion at this stage the smaller will be the drops and the better the emulsion. This energy may be supplied by whirling paddles, high pressure pumps, or vibrating blades as will be shown later but in all the methods the principle is the same. The object being to shear the oil, tear it apart, and smash it into drops. However, some other stage in the preparation is necessary otherwise these drops, no matter how fine, will quickly re-unite until the original quantity of oil is one homogeneous mass again.

When water drips from a tap the globules are at first pear-shaped then as they break away from the metal they very quickly become spherical. This phenomenon is caused by a force called surface tension acting in the skin of the drop, pulling it into a shape which exposes as little of the water as possible to the atmosphere. The body with the smallest surface area in relation to its volume is the sphere—it is well known that the hedgehog curls itself up into a ball when danger threatens, to present as little of its surface as possible to the enemy. Surface tension can be regarded as doing the same to the droplet. Thus, after the first drops of oil have been made there are surface tension forces acting in the skin of each drop which try to prevent further splitting and it becomes more and more difficult to break the drops up into small enough particles to form an emulsion. If, however, some substance could be used which would destroy or at least reduce the strength of the surface tension in the oil drops then it would be easy to split the oil drops still further until they were of such a size as to be capable of forming an emulsion. This is one of the roles of the emulsifying agent, it reduces surface tension and makes droplet formation easier. The other function of the emulsifying agent is to prevent the droplets re-uniting. There are two parts to the molecule of the emulsifying agent; a hydrophilic (water-loving) portion and an oleophilic (oil-loving) portion. The oleophilic 'heads' attach themselves to the oil droplets leaving the hydrophilic 'tails' projecting into the water phase. In this way each drop is surrounded by a layer of emulsifying agent which acts as a buffer and when the droplets collide, as they do many times each second, they are prevented from re-uniting. Emulsifying agents belong to the class of chemical compounds known as surface-active agents because of their ability to bring about these special effects at the surface of liquids. The emulsifier therefore (1) reduces surface tension, (2) stabilizes the emulsion. In addition it must be inert to any chemicals which may be added to the emulsion and must cause no damage to the jute. Most modern surface-active agents are extremely powerful and usually only about 1 part of emulsifier to 20 or 30 parts of oil are required to form a jute batching emulsion.

DEFECTS IN EMULSIONS

The most glaring defect in an emulsion is for it to have the wrong proportions of oil and water. This is so obviously due to carelessness in preparation that no more will be said about it at this stage. Apart from this obvious fault, two defects may arise.

(1) *Creaming.* When an emulsion is prepared it is impossible to make all the drops exactly the same size, some will be much smaller than others and there will be a few quite large drops. In general, the smaller the drops and the less scatter there is in their diameters the

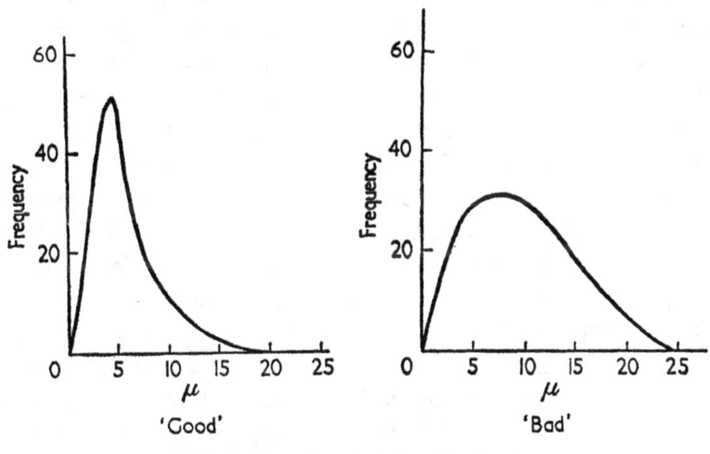

Figure 4.1. Droplet size distributions

better is the emulsion. Figure 4.1 illustrates a 'good' and a 'bad' distribution of droplet sizes. If there are a number of comparatively large drops of oil they will slowly rise to the top of the emulsion because of their lower specific gravity until a layer of them forms at the surface of the emulsion. In emulsion technology this is known as 'creaming'. The same phenomenon will be seen if a bottle of milk is left to stand; the large drops of milk fats rise to the top. As would be expected those emulsions with a large number of big drops will cream more quickly than those with small drops and, again, emulsions made from oils with low specific gravities cream more readily than those with higher gravities.

While creaming is a defect it is not a serious one. Rather, rapid creaming should be taken as a sign of a poor emulsion and attempts should be made to decrease the droplet size. The danger with a creamed emulsion is that supplies of emulsion for the spreaders may be drawn off the top layers which have become heavily loaded with the oil. When this happens the oil content of the jute will be high, but when the emulsion level has dropped and it is now being taken from the oil-deficient layers then the oil content will be low. This trouble can be overcome by arranging a slow-running paddle to keep the contents of all emulsion storage tanks in gentle motion as creaming will only occur in a standing emulsion.

(2) *Breaking*. Breaking can be regarded as the opposite of emulsification where the droplets of the internal oil phase unite to form large drops which then float to the surface of the emulsion. It is a sign of complete instability in the emulsion and once begun cannot be arrested. No amount of re-agitation will split these drops once they have formed and a broken emulsion is useless. The process may be quick or it may take several days, but in jute batching emulsions the presence of drops of free oil on the surface should lead one to suspect a poor emulsion on the point of breaking. Apart from the fact that if this kind of emulsion is put on to the jute the oil droplets will be large and will not spread evenly along the fibres, there will be parts of the emulsion which are deficient in oil and so the oil content of the jute will vary over a period of time. An emulsion may be broken by prolonged violent agitation where the turbulent action breaks down the protective sheath of surface active agent, allowing the drops to coalesce. (This is the basis of butter-making; milk is an emulsion of fat in water and when it is churned the fat droplets conglomerate into butter.) Therefore, while violent mechanical action is necessary and desirable when the emulsion is being prepared it is undesirable and harmful once the emulsion has been made. In storage tanks a gentle stirring action is wanted to prevent creaming but a violent action may break down the emulsion altogether.

SPECIAL ADDITIONS TO EMULSIONS

In order to confer certain particular properties upon a yarn some chemical substance may be added to it. This may be done by treating the finished yarn or cloth but there is much to be gained by adding it to the batching emulsion as the cost of impregnating and drying is eliminated. As long as the additive is compatible with the oil/water emulsion, produces no undesirable side-effects in the process, and achieves the necessary goal then its application along with the emulsion should be considered. Such additives do not usually require great changes in emulsification technique but they do need accuracy in their use and care in making the emulsion. An example of this technique is the addition of rot-proofer at batching. Several rot-proofers are available but one which is commonly used is lauryl pentachlorophenate (LPCP), a brown oily liquid miscible with oil. To make up the emulsion the oil and LPCP are pre-mixed in a special tank and then the emulsion is prepared in the usual way, using this oil/rot-proofer mix just as the mineral oil is used. Amounts of up to 2 or 3 per cent of the LPCP may be added to the jute so it is advisable to reduce the amount of oil somewhat otherwise there will be an excessive quantity of 'oily' material present.

Another example of an addition to the emulsion is found in the use of dyestuffs to give a tint to the fibre. This process is not dyeing, be it noted, and the depth, uniformity, and range of colours obtainable with this technique cannot match those resulting from normal dyeing methods, but for many purposes the tint is satisfactory. The dyestuff may be added via the oil or the water; in the former case an oil-dispersing dye must be chosen which will enter into the oil droplets and be carried along with them to be deposited on the fibre, in the latter case a water-dispersing dye must be used.

EMULSIFICATION EQUIPMENT

The formation of an emulsion is a two-stage process; the correct conditions must be created by the physico-chemical action of the emulsifying agent and the energy of emulsification must be supplied mechanically. Emulsification equipment is not complicated, remembering that the essence of the system is to tear the internal phase into drops and that the more vigorous the action the better is the equipment.

(1) *Paddle-mixers and agitators.* The simplest type of emulsifying plant and one that is common in the jute industry consists of a tank with a rotating paddle inside it (Figure 4.2). There are usually three tanks situated above the mixing tank so that their contents can be

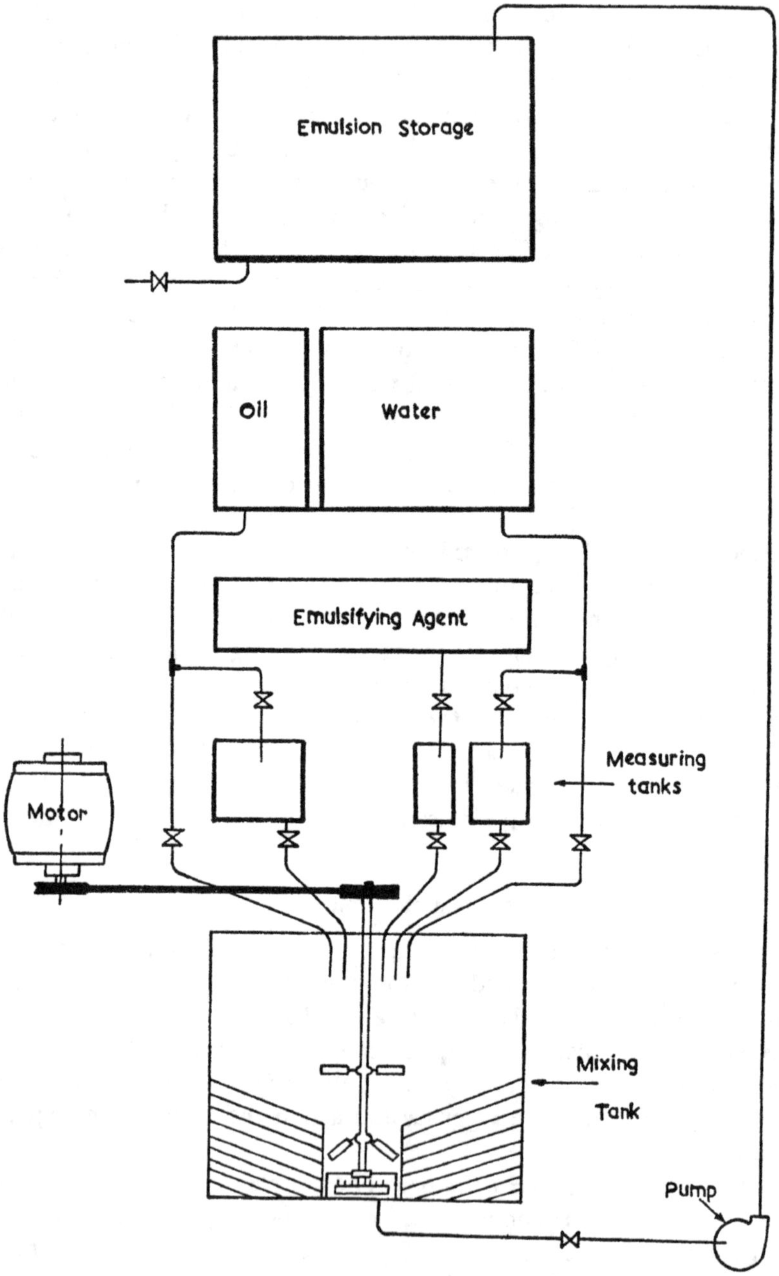

Figure 4.2. Simple emulsion mixing plant

fed to it by gravity. The dimensions of the tanks are in direct proportion to the amounts needed for the emulsion; one holds the water, another the oil, and the third the emulsifying agent. To stop the contents of the mixing tank swirling as a mass when the paddle is running, baffles must be fitted to sides of the tank. These break up the motion of the liquids and give the shearing action which is so necessary. It is good practice to have a small well in the foot of the mixing tank with the agitator blades projecting into it so that the small quantities of emulsifier and water which are added first are mixed efficiently. From the mixing tank the emulsion is pumped to a storage tank where it is held until needed.

This type of equipment will prepare emulsions which are adequate for jute batching but, nevertheless, the emulsions are not of a very high standard. Where an emulsion does not have to remain stable for long periods of time there is no need to install highly efficient emulsifying plant, and as most jute emulsions are used within a few hours of mixing this simple equipment is satisfactory. Even so, some manufacturers adopt the view that good emulsifying machinery is not too expensive and is worth installing for the sake of getting correct emulsification with the minimum of mistakes.

(2) *Homogenizers.* There are many types of homogenizers on the market, all of which work on the same general principle. A coarse mixture of the liquids to be emulsified is forced through a small aperture under high pressure. There are two parts to the homogenizing unit; the pump which generates the high pressures (1,000 lb/in^2 or more) and a special valve with a clearance of a few thousandths of an inch. The degree of emulsification can be controlled by altering the pump pressure or varying the size of the valve clearance or both. Homogenizers can be fitted into a plant preparing emulsion on a batch system. (Note: the term *batch system* is used here in its wider context of one quantity of material being made and passed to storage, then another quantity being made and so on. It does not refer specifically to jute 'batching'.) Figure 4.3 shows a suitable arrangement for using a homogenizer. In the emulsion preparation there are three phases: first, the coarse emulsion is made up in the pre-mix tank; second, this emulsion is pumped under pressure through the homogenizer; and, third, the emulsion is passed to storage.

(3) *Colloid Mills.* These machines are capable of producing extremely fine droplets and, like the homogenizers, they usually work on a coarse pre-mixed emulsion. Basically, the machine consists of a

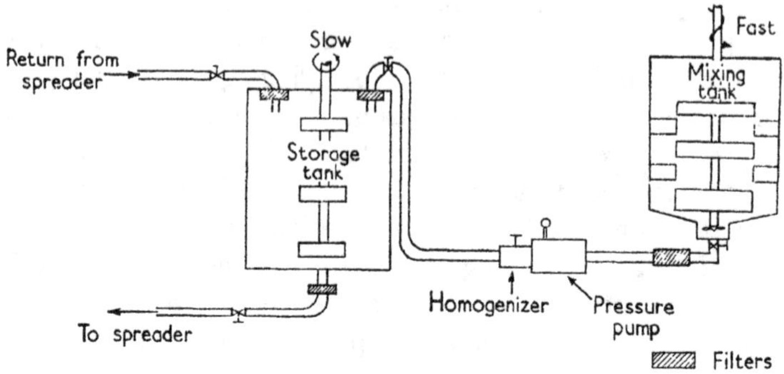

Figure 4.3. Homogenization plant for emulsions

high-speed rotating disk or cone fitting closely inside a shield. The liquid passes between the disk and the shield and in so doing is subjected to strong shearing forces which reduce the particle size. Colloid mills may carry out the work of emulsification or reduce the droplet size of a coarse pre-mixed emulsion.

(4) *Ultrasonic Emulsification.* A plant based on ultrasonic emulsification has been in use in the industry for a few years. The principle is analogous to that used in woodwind musical instruments; in these, air from the mouth is blown across a thin reed causing it to vibrate. The vibration produce air waves which the ear interprets as sound. Ultrasonic vibrations are similar pressure waves but with a frequency too great for the ear to detect. In the ultrasonic homogenizer a jet of liquid strikes the edge of a thin blade, setting up vibrations of the order of 22,000 c/s. These extremely rapid vibrations cause miniature 'explosions' within the liquid, tearing it into fine drops. Figure 4.4 shows the layout of such an emulsifier with the jet and the blade enclosed in a resonating bell to intensify the vibrations. This is the basis of the plant manufactured by Douglas Fraser Ltd, an outline of which is shown in Figure 4.5. The oil, water, and emulsifying agent

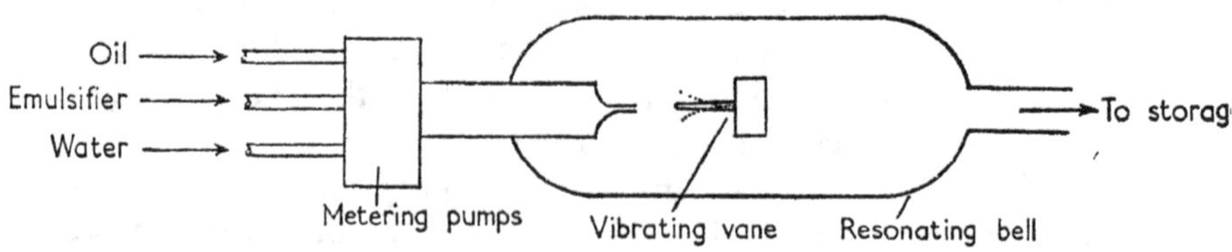

Figure 4.4. Basic principle of ultrasonic emulsification unit

flow by gravity to tanks with ball valves, the valves stopping the flow of liquids when the plant is closed down and switching on a warning light if there is a stoppage anywhere in the supply line. The liquids are withdrawn from these tanks by pumps which are set to

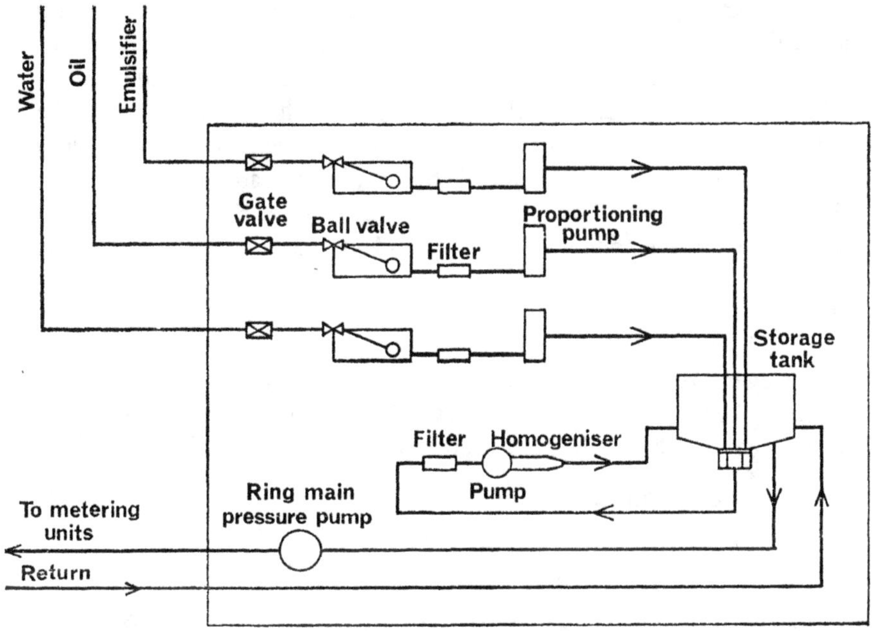

Figure 4.5. Fraser plant

deliver a known volume at each stroke. From the pumps the liquids pass to the ultrasonic homogenizer where the reciprocating blade instantaneously emulsifies the oil. This plant can either be set to produce the required concentration of oil : water ready for application or it may be set to give a 50 : 50 emulsion which then is pumped to a metering unit at the spreader or softener. At the metering unit there are two precalibrated valves, one for metering the 50 : 50 emulsion and the other for metering water. Each valve is set to deliver the necessary quantities within a certain time and, with this system, changes in the oil concentration of the emulsion can be achieved quite easily by adjusting the rate of flow of the water. Once the stock emulsion has been diluted this emulsion is fed to the spreader or the softener. The Fraser unit can emulsify 1,200 lb of oil per hour, enough to add 5 per cent of oil to 24,000 lb of jute in an hour.

THE FORMULATION AND MIXING OF EMULSIONS

If an emulsion with a given oil : water ratio is added to the jute at a certain application rate it is important to appreciate that this will add a fixed amount of oil to the fibre. Emulsion recipe, application, and addition are rigidly interrelated and it is impossible to alter one and leave the other two as they were. If one wishes to change the amount of oil added to the fibre but to keep the amount of water that one is adding the same, then both the recipe of the emulsion *and* the application rate must be altered. If, in the emulsion, there are w parts of water and m parts of oil, and if this emulsion is added to the jute at the ratio of e parts of emulsion to 1 part of jute, then there will be we parts of water added to the fibre and me parts of oil added to the fibre. For example, if an emulsion is made up of 30 lb of oil and 90 lb of water and is to be added at a rate of 20 per cent to the jute, how much oil will be added to the jute?

For every 100 lb of jute, 20 lb of emulsion will be added.

The 20 lb will have oil and water in the ratio of 30 : 90 or 1 : 3. That is to say, there will be $\frac{3}{4} \times 20$ lb of water and $\frac{1}{4} \times 20$ lb of oil

Amount of oil added = 5 lb on every 100 lb of jute

i.e. $$\frac{5}{100} \times 100 = 5 \text{ per cent}$$

Similarly, 15 per cent of water will be applied. It is impossible to add a different amount of water and/or oil without changing the emulsion recipe and the application rate.

The common range of application rates for jute is from 16 to 25 per cent, with up to 30 per cent being added when cuttings are being run through. With the usual 5 per cent of oil, on the finished goods this corresponds to about 70–80 per cent water in the emulsion, for stainless yarns about 92–95 per cent water. The quantity of emulsifying agent depends upon the particular type used but is usually in the ratio of 1 to 20 or 30 parts of oil.

One of the essentials in emulsion preparation is to measure out the ingredients accurately. Unless this is done there will be an unnecessarily large day-to-day variation in the oil content of the yarn and while this is perhaps not of vital importance in 5 per cent oil material it can assume very great importance in stainless goods where the variation may take the oil level over the permitted 1 per cent. In many plants the liquids are measured out from tanks fitted with sight-glasses which have been calibrated from the calculated volume of the tank but it is essential to check the calibration by adding known volumes to the tanks. It is much better practice, however, to fit fluid meters on the supply lines or to have an accurate level indicator in the tank. Fluid metering is, of course, an integral part of the Fraser ultrasonic emulsifying plant and it can be shown that this system gives a lower day-to-day variation in the oil content of the yarn than one relying upon sight-glasses and dipsticks to measure out volumes. Accuracy is particularly important when additional substances are being added to the emulsion where too little additive may fail a consignment of yarn when it is tested or too much may make the process uneconomic. Needless to say, all the ingredients of an emulsion should be added by weight and not by volume.

As most jute batching emulsions are made with the simple stirrer-type apparatus this method of preparation will be dealt with at length. With homogenizers or colloid mills it is common practice to make a coarse emulsion in a stirrer-plant and then reduce the particle size later; with the ultrasonic method the process is automatic and all that is necessary is to set the pumps and valves correctly. The method used for the simple plant is known as the 'mayonnaise' method. The emulsifying agent and an equal amount of water are added to the tank and stirred together. As this may only amount to 2–3 gal of liquid it is clear that a sump in the bottom of the tank helps to mix the two liquids at this stage. The oil is added slowly and soon a thick creamy paste forms, the 'emulsion base', which has the consistency and appearance of mayonnaise—hence the name of the method of mixing. This base has powerful oil-absorbing properties and is expanded by adding the remainder of the oil. While the oil is forming the emulsion base the stirrer is breaking it up into small drops which acquire a protective coating of the surface active emulsifying agent. The water is then added slowly, the stirrer operating continually. Once about half the water has been added at the slow rate the remainder can be put in quickly. After a few minutes final mixing the emulsion is pumped to a storage tank where it is stirred gently until required. Some manufacturers prefer to heat their emulsions, in which case heating may be carried out by passing the emulsion through a heat-exchanger on its way to the storage tank, heating it once it is in storage by means of closed steam coils or heating while it is on its way from the storage tank to the point of application.

CHAPTER FIVE

Jute Batching

THE sequence of operations at the start of the jute spinning process depends upon which class of yarns is being made. For the better grades, such as those destined for hessian fabrics, where the raw material is long jute from which the root end has been cut, the principal machine is the spreader, but for the poorer grades of yarn the jute is passed over the softener as the short nature of the raw material precludes the use of the spreader. Because of this division the two systems will be treated separately, but before doing so it may be advantageous to discuss certain terms which are common to both. It has already been indicated that the term 'batching' strictly refers to the addition of oil and water to the jute, but the use of the term has spread to associated features at this stage in the process. The department where the jute is taken from the bale and prepared for carding is called the 'batching-house'. As will be explained shortly a blend of different types of jute is made up to suit the particular class of yarns being spun, this blend being known as the 'batch'. 'Conditioning' or 'maturing' refer to the resting stage which jute is given after the water and oil have been applied, it lasts longer with low-grade batches to allow the hard, barky, root material to become softened before passing on to the cards.

Since jute is a product of nature, and as such is subject to the vagaries of soil and climate, its properties are by no means constant at all times. If only one strain of jute were used until that was exhausted, then another type fed into the process, then a few months later yet another type fed in, it is obvious that there would be a continual change in the strength, colour, and regularity of the yarn from month to month. If, on the other hand, the different kinds of jute which are available are thoroughly mixed together into one homogeneous lot then this will provide a supply of raw material which is reasonably constant and which will spin a yarn of a suitable quality at all times. Certain factors must be borne in mind when the grades of jute are being selected to form a batch. It is better to avoid large differences in the physical properties of the grades being blended. For instance, it is not good practice to blend a high quality jute with a low quality one, since the good qualities of the former will be completely swamped by the poor qualities of the latter. For this reason blending is confined to similar grades of jute.

Commonly two to six grades of long jute are put into a batch for hessian-type yarns; for sacking yarns, cuttings, low-grade long jute, and mill wastes are used. It is desirable to express the components in terms of their relative percentages in the batch, for example.

Quality	*Quantity*	*Percentage*
Mill Hearts	2 bales	18
Export Lightnings	4	36
Grade Tossa 4	4	36
Export Hearts	1	9
Northern Tossa X-Bottoms	1,800 lb	30
Northern White C-Bottoms	1,800	30
Cuttings	1,200	20
Habi-jabi	900	15
Bale ropes	300	5

Examples are given below of the types of jute which may be used for the various classes of yarns.

Fine yarns (3½–6 lb/sp)	Top quality Dacca Tossa 4s and 5s, Crack Hearts, or similar grades.
Medium yarns (6–20 lb/sp)	Medium quality, Mill class of white jute or Export white, Grade Tossa, Out-port Tossa 2/3s, 4s. Warp batches always higher than weft. Kenaf may be included in weft batches.
Sacking yarns (12–40 lb/sp)	Warp from low marks of long jute, weft from cuttings, tangles, bale ropes, thread waste, and some low-mark long jute.

HESSIAN BATCHING

The various marks of jute are assembled at the start of the processing line. The ropes round the bales are cut through with an axe and laid aside to be processed separately. Under the extremely high pressure of the pucca baling press the jute becomes hard and as stiff as wood. Before the fibre can be handled satisfactorily it must be made more flexible. This is done by a machine called a bale-opener. The bale-opener is a massively built machine with heavy fluted rollers, intermeshing with each other. Figure 5.1 shows two types which are in

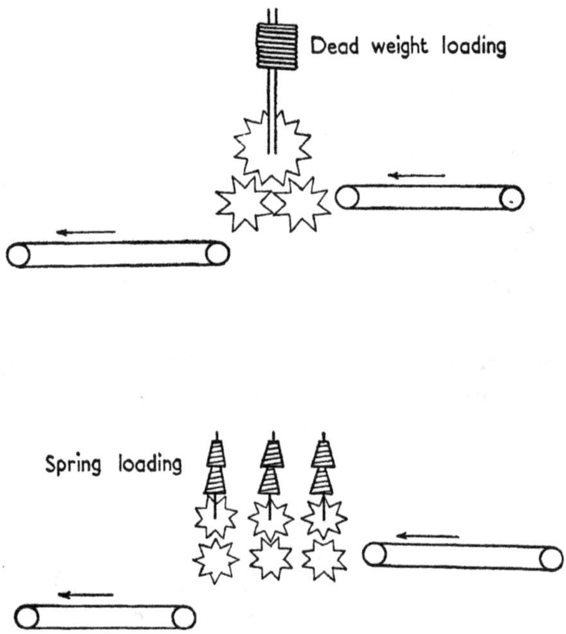

Figure 5.1. Two types of bale-opener

common use. When the ropes have been cut off, the bale still retains its rectangular shape and the jute is pulled off by hand in slabs or heads, each head comprising a bundle of reeds which have been loosely twisted together weighing 8–9 lb. Complete heads are fed into the bale-opener where the action of passing between the fluted rollers under pressure flexes the jute and it emerges from the machine soft and pliable. Most bale-openers operate at 28–30 ft/min and can handle 1 bale in about 2 min.

Recently an automatic bale-opening range has been developed by Douglas Fraser Ltd with a view to saving labour at this stage. The bale is placed on a feed lattice and is carried up into the bale-opener beneath rotating knives which cut the ropes. The bale is then squeezed by three pairs of heavy fluted rollers which soften and open out the heads into the form they were in before baling. The bale is then discharged on to a special trolley. Using this machine one man can handle 30 bales per hour.

The next step is to split the heads up into smaller bundles, called 'stricks', for feeding to the spreader. The heads are untwisted by hand, split lengthways into stricks weighing 3–5 lb. The stricks are then given a half twist at their middle, folded, and placed neatly on a barrow. The stricks should be as nearly the same size as possible and striking-up, as the operation is called, is a matter of experience. The first stages of blending begin at the bale-opener, for a head is taken from each mark of jute in turn and fed through the machine so that the pile of jute at the delivery end, from which the strikers-up work, is a mixture of the different marks. As the barrow is built up with stricks from the various marks in the batch further mixing and blending goes on. Once the barrow is full it is pushed to a holding-area to supply the spreader feeds.

THE SPREADER

The jute spreader was developed from the earlier Good's machine of the hard fibre trade and has now supplanted the softener for hessian-type yarn manufacture. Figure 5.2 shows the essential points of the machine.

The stricks are taken off the striking-up barrow one by one and laid by hand on the feed sheet of the spreader, the root end of one strick overlapping the crop end of the previous one. This is the point where the separate and individual reeds of jute are assembled into a continuous sliver. The stricks pass between a pair of fluted feed rollers and on to the pins of the slow-moving pinned lattice known as the slow chain; above the slow chain there are two or three lantern rollers to press the jute firmly down on to the pins. As may be imagined, the construction of the pins (and indeed the whole machine) is rugged. Halfway along the machine the material is transferred from the pins of the slow chain to those of a similar chain having a higher surface speed. Because of the greater linear speed of the fast chain the jute

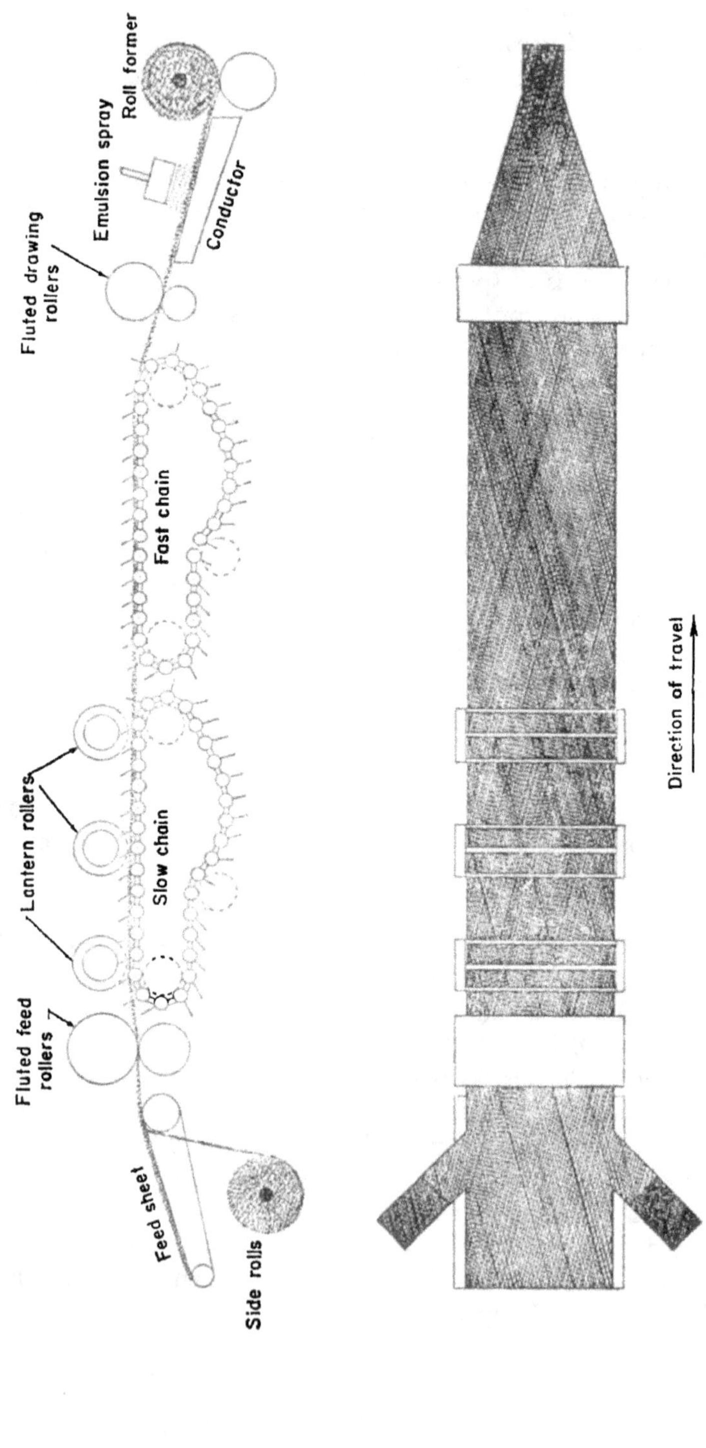

Figure 5.2. Jute spreader

is combed and drawn out, i.e. drafted, at this transfer point. When a fresh strick passes between the feed rollers and on to the slow chain it can be clearly seen that the root end is much heavier and bulkier than the remainder and as it comes under the action of the fast chain the faster moving pins tear and comb out this root end while the rest of the strick is securely held by the slow chain pins. This action continues until so much material has been transferred from the slow chain to the fast one that there is no longer sufficient jute imbedded in the slow chain to hold the strick back and it suddenly whips through the slow chain pins. This phenomenon of sudden release of restraint and its associated rapid fibre movement is met with at other stages in the process and is known as 'gulping'. Wherever it occurs it is undesirable since it means that the tail-end part of the material has not had the full treatment it needed.

The main spreader draft operates between the slow and fast chains and the linear speed of the latter divided by that of the former gives the draft at this point. If there is a draft of, say, 6 then 1 ft of material on the slow chain will become 6 ft on the fast chain. Concurrent with this attenuation or drawing-out in length there is a reduction in the sliver count.

The basic equations concerning draft are

$$\text{(1) Machine draft} = \frac{\text{Greater linear speed}}{\text{Lesser linear speed}}$$

$$\text{(2) Length fed} \times \text{draft} = \text{Length delivered}$$

$$\text{(3) } \frac{\text{Count fed}}{\text{Draft}} = \text{Count delivered}$$

The jute comes off the fast chain, passes between a pair of fluted delivery rollers and is guided down an open-topped channel where the emulsion is added, either by a pressure spray or by a gravity-fed drip weir. The final action is to collect the sliver in a form suitable for the next stage of carding.

There are four factors on the spreader which determine its efficiency from the points of view of quality and production.

(1) The fibre must be fed into the machine as evenly as possible.
(2) The stricks must be combined into a continuous sliver which is then drafted to the correct count—this action to be combined with a certain amount of preliminary splitting and opening of the stricks.

(3) The emulsion must be applied uniformly and at the proper rate.
(4) The delivered sliver must be in a state suitable for feeding to the breaker card.

FIBRE FEED

The spreader feeder is presented with a barrow containing perhaps 200 stricks and weighing 1,000 lb, enough material to last about half an hour and his problem is to feed all that jute to the spreader at the same rate from start to finish of the barrow and do the same for the next barrow and the next, hour after hour during the day. Without some assistance it would be extremely difficult for him to do this consistently and this assistance is given at the spreader by a weighing machine with an additional pointer on it. The barrow is pushed on to the platform of the weighing machine which is situated conveniently to the spreader feed and the dial of the balance registers the weight of jute on the barrow (after allowing for the tare of the barrow). If now the jute is taken off, one strick at a time then the reading on the weighing machine scale will gradually fall until it registers zero when the barrow is empty. In this way there is a ready means of knowing how much jute has been fed to the spreader but as yet no account has been taken of how quickly it has been fed. Unless the jute is fed at the correct rate in terms of pounds per minute it will be impossible to achieve the correct count of sliver. If the spreader feeder was given a clock and told that he must empty the barrow in, say, 32 min then he could judge fairly accurately that he would have to feed between 150 and 160 lb of jute every 5 min to achieve a regular feeding rate. Apart from the obvious disadvantages of providing clocks for each spreader this method would be liable to error for whenever the spreader stopped for any reason the feeder would almost certainly fail to note the exact time and so find it difficult to pick up the proper feed rate immediately. This difficulty can be overcome if a 'clock' driven by the machine itself is provided. The feeder can then be told that he must feed a specific weight of jute to the machine in one complete revolution of the machine-driven clock pointer or some fraction of it. This, in fact, has been commonplace on jute machinery for many years and the term 'clock length' is used. The clock length is simply the distance moved by the feed sheet in one revolution of the clock pointer. The weight of jute which is fed on to this length of feed sheet is called the 'dollop'.

Thus if the dollop is 700 lb and the clock length is 15 yd then the weight per unit length being fed is

$$\frac{700}{15}\times\frac{100}{1}=46{\cdot}7 \text{ lb/100 yd}$$

On the spreader the machine-driven clock pointer is placed on the front of the weighing machine dial and as the spreader runs it moves slowly round the face of the dial, giving the operative a pace to work to. By means of suitable gearing the clock pointer is made to move around the dial at a speed which will be matched with the dollop weight and the clock length. All the feeder must now do is to remove jute from his barrow at such a rate that the weighing machine pointer and the driven pointer are coincident at all times and he will be certain that he is feeding the jute at the correct rate. On the spreader this driven pointer is often called the 'slave' pointer but this is a misnomer; the driven pointer demands that the machine will be fed at a certain rate and therefore *it* is the master and the weighing machine pointer, which must follow it, is the slave.

Figure 5.3 illustrates the gear drive to the driven pointer on a spreader, the motion being taken off the feed rollers. In the previous paragraph it was said that the clock length was equal to the total

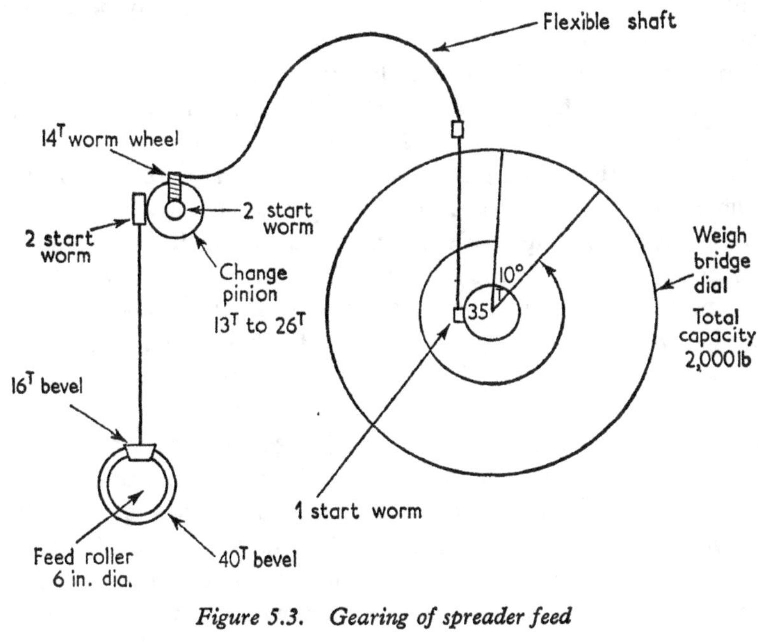

Figure 5.3. Gearing of spreader feed

distance moved by the feed sheet while the driven pointer moves through one revolution. For the spreader this is not quite true since, because of machine design, the pointer only travels through 350 degrees and not 360 degrees. The clock length is calculated by assuming that the feed to the machine is driven by the pointer (in fact, of course, it is the other way round) and finding the total distance the feed sheet moves. To enable quick changes to be made to the clock length there is a clock length change pinion in each gear train. In the example shown in Figure 5.3 the clock length is

$$\frac{350}{360}\times\frac{35}{1}\times\frac{14}{2}\times\frac{\text{change pinion}}{2}\times\frac{16}{40}\times\frac{6\times 3{\cdot}142}{36}\text{ yd}$$

$$= 25\times\text{change pinion yd}$$

i.e. the gearing constant is 25 and Clock length $= 25\times$ change pinion yd

Since the change pinions in the gear drive to the pointer are in the range 13–26 teeth, alterations in the clock length of the order of 5–10 per cent can be made. This is of advantage if changes in the feeding rate are wanted or if the main draft changes on the machine are too coarse to effect the necessary alteration in sliver count. By using the spreader feed change pinion in conjunction with the main draft change pinion a wide range of operating conditions may be achieved, as will be seen later.

While this system of a master driven pointer and a slave weighing machine pointer offers a convenient and simple means of regulating the spreader feed it may not achieve its object under certain conditions. Ideally, the weighing machine dial should be directly in front of the feeder and he should look straight at it. However, in many installations this is impossible due to the layout of the work-place and the space available and the weighing machine dial is a few feet away from the feeder and situated at an angle to his line of vision. If this is so then there are certain positions on the dial where, to the feeder, the two pointers appear to be exactly in line, but when viewed from directly in front of the dial, as they should be, they are seen to be separated and the weighing machine pointer will either be lagging behind or leading the driven pointer. Under these conditions the feed will not be constant from start to finish of the barrow and one often finds that as the barrow is emptied the feeding rates gradually become greater, simply because of this positioning error resulting from parallax when the feeder views the two pointers. An attachment to the

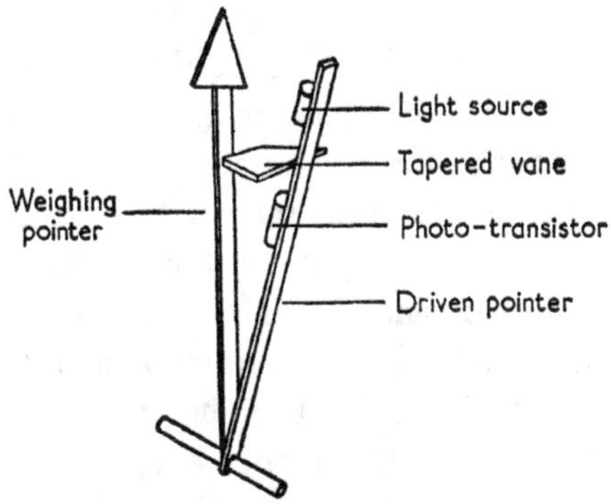

Figure 5.4. Essential features of the B.J.T.R.A. feed regularity meter

spreader feed called the 'Feed Regularity Indicator' has been developed by the B.J.T.R.A. to overcome this difficulty, Figure 5.4 showing the essentials of the method. A light source and a photo-transistor are fixed to the driven pointer a few inches apart and a triangular-shaped metal vane is attached to the weighing pointer. The vane can interrupt the beam of light shining from the light source on to the photo-cell by passing between them and, since it is tapered, the amount of light cut off becomes proportionately greater as the vane passes further between the two. In this manner the amount of light cut off can be used to measure how far the vane has penetrated between the light and the photo-transistor. The light falling on the cell generates a small

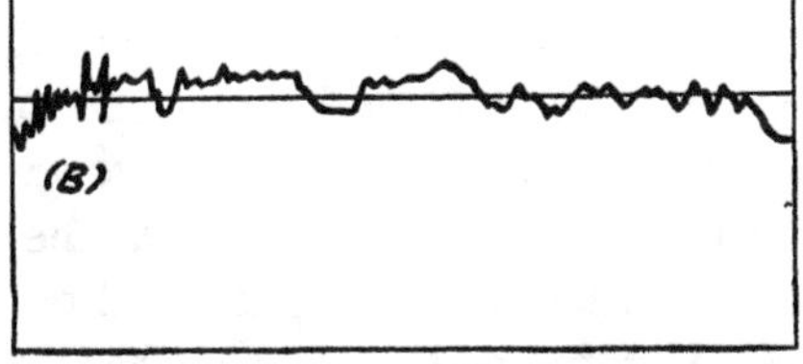

Figure 5.5. Records of pointer separation: A, without Feed Regularity Indicator; B, with Feed Regularity Indicator

current of electricity which is used to operate an indicator at the front of the machine directly above the feed sheet and in clear view of the operative. When the two pointers are exactly in line, as they should always be for perfect feeding, a certain amount of current is generated and the indicator registers 'correct rate of feed' but if the weighing pointer lags behind the driven pointer then more light is let past, more current is generated, and the indicator shows 'feed too light'. On the other hand, if the weighing pointer gets ahead of the driven one then the indicator registers 'feed too heavy'. Figure 5.5 shows two records of the amount by which the weighing pointer has been separated from the driven pointer, one being taken with the Feed Regularity Indicator in use and the other when it was not; the improved uniformity in the rate of feeding can be seen. The advantages of the Indicator are

(1) It assists the spreader feeder to keep a uniform rate of feed.
(2) Being situated directly above the feed sheet it is facing the feeder and there is no parallax.
(3) The Indicator shows clearly if the feed is heavy or light.

The Indicator can, if necessary, be coupled to a pair of counters which record the total length of sliver produced by the spreader and the amount of sliver produced within certain prescribed limits. This is done by arranging that one counter will operate as long as the spreader runs and the other will operate as long as the output current of the photo-cell lies within a certain range but to stop as soon as the current exceeds it. In other words, as long as the weighing pointer follows the driven pointer within a certain tolerance then both counters record, but whenever the feed is excessively heavy or light then one of the counters will stop. This can form the basis of a quality assessment for spreader feeding, e.g. if the limits of separation which will be tolerated are ± 10 lb then both counters will operate as long as the weighing pointer is within 10 lb of the driven one. But as soon as the pointer exceeds the tolerance one of the counters will cease recording until the weighing pointer returns within the 10 lb limit. The total length of sliver put out by the machine may be 16,000 yd in a day and there may be 14,500 yd of 'good', i.e. within-limit, sliver recorded; then one may say that for this period the spreader was being fed for 91 per cent of the time in a satisfactory manner.

Before leaving the feeding arrangements on the spreader it is necessary to discuss 'leader' rolls. These are two rolls of spreader sliver which are brought back to the feed end of the machine where they are entered through two special channels at the top of the feed sheet and pass into the feed rollers along with the hand-fed stricks. The purpose of using leader rolls is to give a more uniform sliver with a cleaner, neater edge to the roll of delivered sliver but off-setting these advantages is the occasional trouble experienced when the sliver coming from them breaks or becomes tangled, and the small amount of extra labour required to bring them from the delivery end of the machine to the feed. The leader rolls may be drawn from the normal supply of sliver rolls, in which case some of the jute gets an additional treatment with emulsion, or alternatively a supply of 'dry' rolls can be made specially for leaders—the latter method is to be preferred. The use of leaders is optional, some manufacturers being of the opinion that the rolls of spreader sliver are satisfactory without them.

DRAFTING

Most of the drafting on the spreader occurs at the transfer from the slow chain to the fast chain. Although small drafts—usually referred to as leads—are present at every transfer point, i.e. feed sheets to feed rollers, feed rollers to slow chain, fast chain to delivery rollers, delivery rollers to roll former. The object of a lead is to keep the material taut as control is passed from one stage to the next.

Typical values for speeds, draft, and leads are

Feed sheet	6 yd/min	
Feed rollers	7·3	Lead 21·5 per cent $\left(\frac{7{\cdot}3-6{\cdot}0}{6{\cdot}0}\times 100\right)$
Slow chain	8·0	Lead 8·8 per cent $\left(\frac{8{\cdot}0-7{\cdot}3}{7{\cdot}3}\times 100\right)$
Fast chain	44·8	Draft 5·6
Delivery rollers	58·2	Lead 29·8 per cent
Roll former	64·0	Lead 10·0 per cent

Overall draft 10·67 $\left(\frac{64{\cdot}0}{6{\cdot}0}\right)$

For each spreader there is a draft constant derived from the gearing; this is a numerical constant calculated from the number of teeth in the pinions of the gear train and the surface speeds of the feed and delivery rollers. The draft on the spreader can be altered to suit the production requirements by changing one pinion in the gear train. This pinion is called the draft change pinion; the pinion needed for a certain draft is found by dividing the draft constant by the draft, e.g.

Draft constant: 400

Draft required: 11

Draft pinion: $\frac{400}{11} = 36{\cdot}4$ teeth

Since pinions cannot have fractions of a tooth,

Draft pinion used: 36 teeth

Actual draft: $\frac{400}{36} = 11{\cdot}1$

Figure 5.6 illustrates a typical gear train for a spreader.

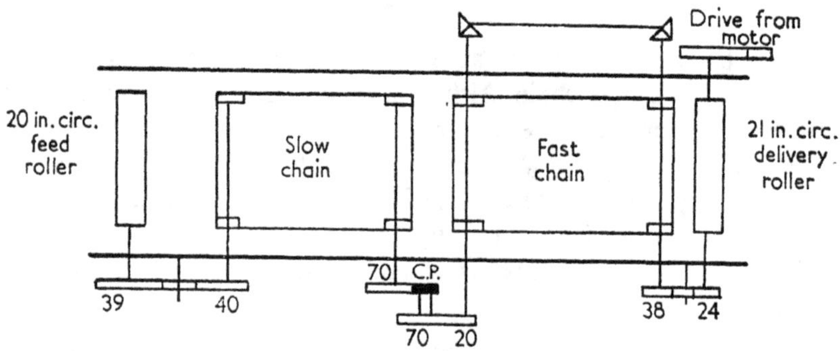

Figure 5.6. Spreader draft gearing

For illustration, the method of calculating the draft constant on the spreader will be shown. Assume that the delivery is driven from the feed and find the number of yards delivered as a result of 1 yd being fed in at the feed end. In the gearing of Figure 5.6 1 yd of feed requires $\frac{36}{20}$ revolutions of the feed roller and therefore the length of sliver delivered when the feed roller rotates through $\frac{36}{20}$ revolutions is,

$$\frac{36}{20} \times \frac{39}{30} \times \frac{70}{\text{change pinion}} \times \frac{70}{20} \times \frac{38}{24} \times \frac{21}{36} = \frac{397}{\text{change pinion}}$$

i.e. Draft constant = 397

By means of a range of draft pinions, different drafts can be selected to produce the count of sliver required. For the gearing illustrated in Figure 5.6 these are

Change pinion	*Draft*
33	12·0
36	11·0
40	9·9
44	9·0

It may be, however, that the changes in the draft shown are too large to suit a particular set of circumstances and in this case it is possible to obtain much finer steps of draft by using the clock length changes in conjunction with the spreader drafts. To show the details of the calculation of the delivered sliver count when certain pinions are used in the clock length gearing and the draft gearing, an example will be given.

Dollop weight	1,000 lb
Clock gearing constant	10
Spreader draft constant	350
Clock length change pinion	20
Draft change pinion	35
Emulsion application	25 per cent

Clock length $= 10 \times 20$

$= 200$ yd

Feed sliver count $= \dfrac{1000 \times 100}{200}$ lb/100 yd

$= 500$ lb/100 yd (2,500 ktex)

Emulsion applied $= \dfrac{25}{100} \times 500$ lb on each 100 yd of jute

$= 125$ lb

Total weight fed in $= 500 + 125$

$= 625$ lb/100 yd

Spreader draft $= \dfrac{350}{35}$

$= 10$

Delivered sliver count $= \dfrac{625}{10}$ lb/100 yd

$= 62{\cdot}5$ lb/100 yd (312·5 ktex)

TABLE 5.1. EXTENDED LIST OF SPREADER SLIVER COUNTS USING DRAFT AND CLOCK LENGTH CHANGE PINIONS

Clock change pinions	*Draft change pinions*			
	33	36	39	42
13	61·4	67·0	72·6	78·3
14	57·0	62·3	67·5	72·6
15	53·2	58·1	62·5	67·7
16	49·9	54·4	59·0	63·5
17	—	51·3	55·6	59·8
18	—	—	52·5	56·5

Table 5.1 shows the range of spreader sliver counts which could be prepared from a combination of feed gearing change pinions and draft change pinions, the figures being illustrative only and referring to a 20 per cent application, a 2,000 lb dollop, a feed constant of 25 and a draft constant of 397.

Thus, by a judicious choice of the two pinions it is possible to produce spreader slivers which differ in count by only 1 or 2 per cent. However, there is another matter to be taken into account and this is the rate at which the spreader feeder can actually work and still maintain an even pace for laying the stricks neatly on the feed sheet. Jute spreaders are constant delivery speed machines, i.e. the revolution rate of the delivery rollers is governed by the motor speed and the drive pulley dimensions, and when the draft is changed it is the feed sheet's linear speed that alters. If the draft is increased the feed sheet travels slower and if it is decreased then the feed runs faster.

Besides the general effect of altering the sliver count, drafting plays an important part in determining sliver quality. Spreader sliver is always very variable in count over short lengths, an unavoidable feature of the material and the manner of forming the sliver. If one weighs short lengths of sliver (18 in., for example) and plots the weighings in the order of cutting on a graph, then one can pick out definite wave-like variations in the weights. Figure 5.7 shows the results of such a test.

This is typical of spreader sliver, with short pieces as light as 25 lb/100 yd and others as heavy as 130 lb/100 yd. The peaks of the waves occur at regular intervals which measurement shows are equal to the distance between successive root ends of the stricks on the feed

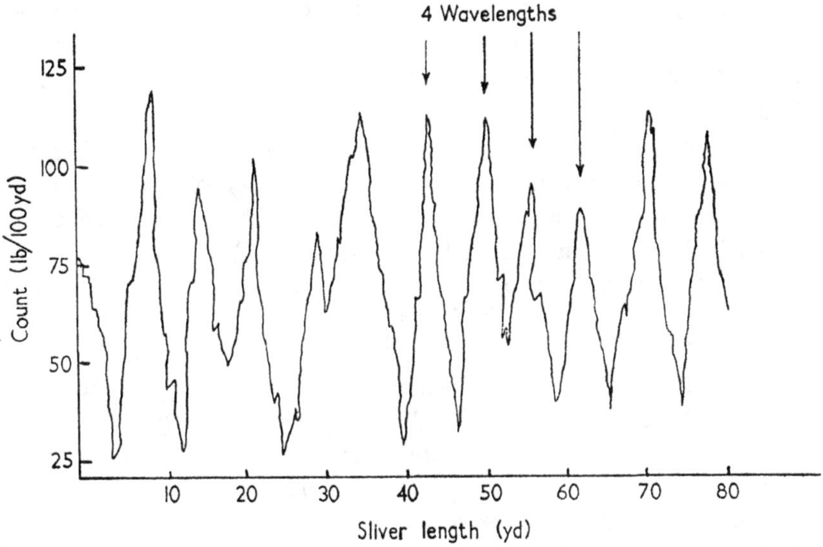

Figure 5.7. Spreader sliver, count of 18 in. lengths

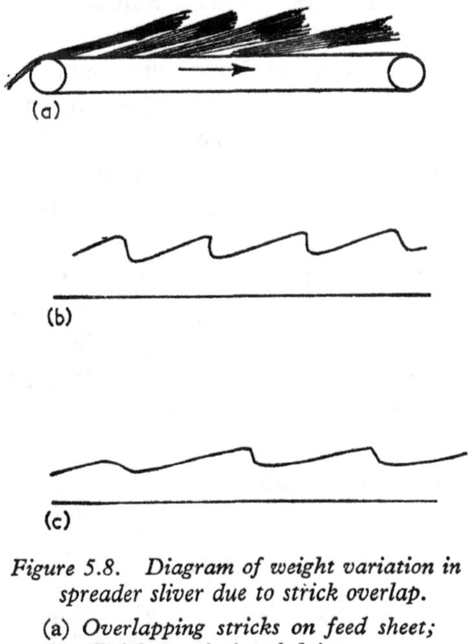

Figure 5.8. Diagram of weight variation in spreader sliver due to strick overlap.

(a) *Overlapping stricks on feed sheet;*
(b) *Weight variation fed in;*
(c) *Weight variation delivered*

sheet multiplied by the spreader draft. The height of the waves depends upon the size of the stricks, being greater for large stricks and smaller for lighter stricks. Figure 5.8 has been drawn to show how this comes about.

As the root ends of one strick overlap the body of the preceding one on the feed sheet it follows that there is a section of material on the feed sheet that is approximately twice the weight of that section immediately before it and immediately after it. When this extra heavy piece is carried forward to the pins of the fast chain then it produces a length of drafted sliver which is still heavier than that before it and after it—all that the draft does is to stretch out the double-weight section. When the delivered sliver is cut into short lengths the count is high at the point where the double-weight portion begins, i.e. at the leading end of each new strick, and falls as the bottom and the upper stricks taper away to their crop ends. Before the count falls to zero, however, a new strick has been thrown on the feed sheet and entered the pins of the two chains, and the weight pattern is repeated. It will readily be seen from Figure 5.8 that the wave-length (the distance from peak to peak) will be greater if the distance between succeeding stricks is great and if the draft is high. Similarly, if heavy stricks are overlapped then there will be a corresponding large amplitude (the height of the wave). These conditions are found when large stricks are used to feed the spreader; because they are heavy they need not be laid on the feed sheet so frequently and therefore the distance between root ends is great. Unless a heavy sliver is taken off the machine, a high draft will be needed to handle the heavy feed and, finally, the large bulky root end will give a wave with a high amplitude. Small stricks, on the other hand, produce a more regular sliver but they do require more labour and effort both at striking-up and the spreader feed. Under normal circumstances the spreader feeder cannot feed much faster than about 10–12 stricks per minute which, for normal rates of feed, requires stricks of about $2\frac{1}{2}$ lb. This represents the minimum strick size, but in practice stricks of 5 lb are common since less labour is required at striking-up.

EMULSION APPLICATION

The emulsion is kept in its storage tank until required and then drawn off by a pump and fed into a ring-main. Figure 5.9 illustrates a typical system of pipelines in the batching department. The ring-main travels

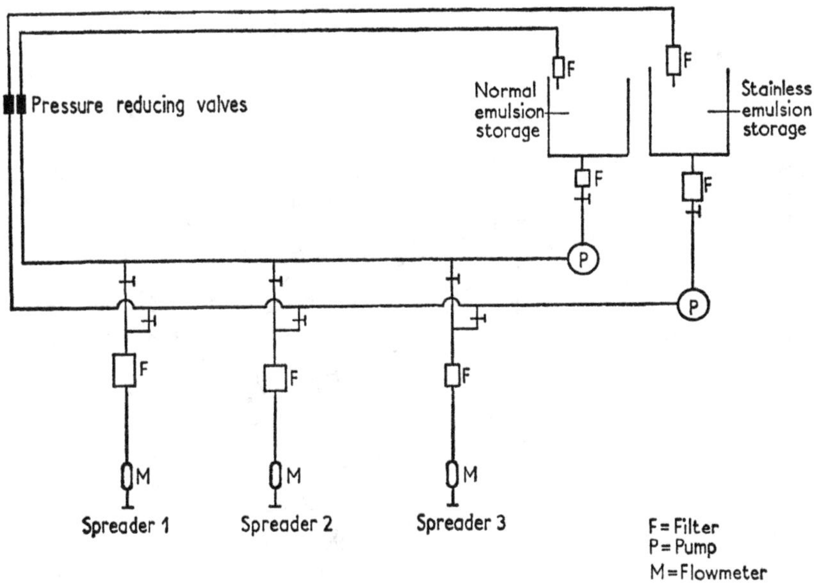

Figure 5.9. Emulsion distribution system

to the spreaders and back to the storage tank, appropriate filters being placed at the exit and the return of the storage tank to keep the liquid as free as possible from dust and dirt. At each spreader a supply line is tapped off, carrying a pressure-reducing valve if necessary, in addition to a throttling valve which cuts off automatically when the spreader is stopped. An alternative supply system may consist of a gravity-feed storage tank which delivers the emulsion to a second smaller storage tank by the spreader from whence it is drawn off by means of a low pressure gear-pump operated by the spreader itself. Whichever method is adopted it must have an adequate series of filters, including one as near the sprays as possible, and an automatic cut-out which will shut off the flow of emulsion whenever the machine stops.

In Chapter 4 the necessity for accuracy in compounding an emulsion and supplying it at the correct rate was emphasized. While it is true that doubling after the spreader helps to even out some of the irregularities of moisture and oil contents, by no means all will be eliminated and, more important, a wrong level of moisture or oil cannot be corrected. For these reasons it is vital that the application be correct.

Figure 5.10 shows the amounts of moisture and oil which will be added to the jute for different application rates and emulsion recipes;

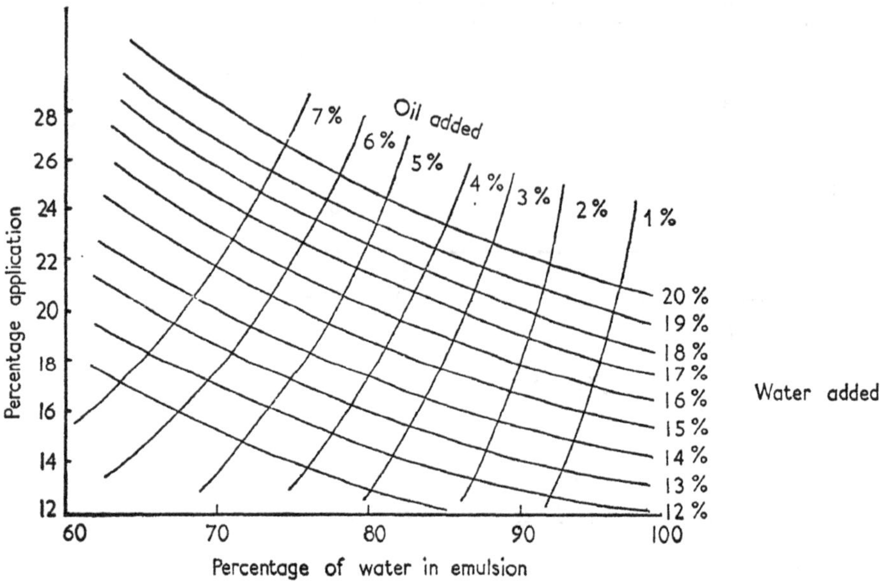

Figure 5.10. Percentages of oil and water added to the jute from various emulsion recipes and applications

notice again that there is one, and only one, combination of application and recipe that will give a specific moisture and oil addition.

The rate at which the emulsion is being fed to the jute may be indicated by one of three methods. The commonest of these is the Bourdon pressure gauge but flowmeters offer a better alternative. Finally, the Fraser ultrasonic plant meters the emulsion by valves and pumps and no external indicator is needed. The Bourdon pressure gauge is of the common type met with in steam-raising plant, water mains, etc., and although it has the advantages of cheapness and robustness, as a means of indicating the rate of flow it has serious drawbacks. Strictly speaking, it is impossible to give an accurate indication of the flow-rate with such a method—it is analogous to measuring the current flowing in an electrical circuit with a voltmeter. The main fault with using a pressure gauge is that when a blockage occurs downstream from the gauge, perhaps at the sprays themselves, then the fluid pressure in that part of the pipe will *rise* and through back-pressure the reading on the gauge will increase, giving the impression that more liquid is passing whereas in fact the flow has been restricted. Before any idea of the amount of emulsion being sprayed on to the jute can be obtained, it is necessary to run calibration tests to find the relationship between gauge pressure and rate of flow; this must be done individually for each spreader, for if there are two or more spreaders operated off the same ring-main then there will be a fluid pressure drop between them and consequently the same gauge pressure at each spreader will not give the same rate of emulsion flow. Another disadvantage in their use is that they are liable to error if the viscosity of the emulsion is changed, as may happen when changing from a stainless emulsion to a 5 per cent oil one. For example, one test showed that when the gauge pressure was kept constant at 16 lb/in^2 and a 20 : 80 oil–water emulsion passed through, then the rate of flow was 6·20 lb/min, but when the emulsion recipe was altered to give 30 : 70 ratio the flow rose to 6·5 lb/min, an increase of 5 per cent. Nevertheless, successful control of the application rate can be achieved with this simple apparatus provided the emulsion is clean, the filters are well maintained, and the flow/pressure calibration is checked frequently.

The most common type of flowmeter met with in the jute industry is the variable orifice type (Figure 5.11). The emulsion flows upwards

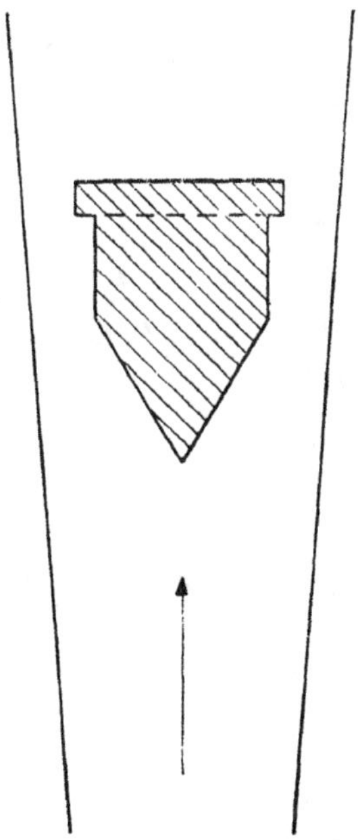

Figure 5.11. Variable orifice flowmeter

through a tapering tube containing a specially shaped metal float. The force of the emulsion passing the float causes it to rise and the height it rises within the tapering tube gives a measure of the amount of emulsion that is passing. The float is held at this height by a balance of forces; the emulsion flow is tending to make it rise but its weight is holding it back. The tube is calibrated in gallons per hour, pounds per minute, litres per minute (or some such convenient unit) for certain operating conditions of pressure, temperature, etc. The simplest form of flowmeter comprises a glass tube with the scale etched on it but more refined types are available which have external indicators capable of operating warning bells or flashing lights if the emulsion flow-rate falls outside certain prescribed limits. These meters are sturdy but quite sensitive to even small changes in the rate of flow and their advantages over pressure gauges are

(1) They register directly the amount of emulsion passing on to the jute; this makes it easy to calculate the percentage application.
(2) Floats can be obtained which are immune to changes in the viscosity of the emulsion and so changes from stainless to 5 per cent material can be made without any adjustment being required to the meter.

The Fraser ultrasonic system has already been dealt with, suffice it to say that the rate of flow is decided by the position of the various valves within the unit.

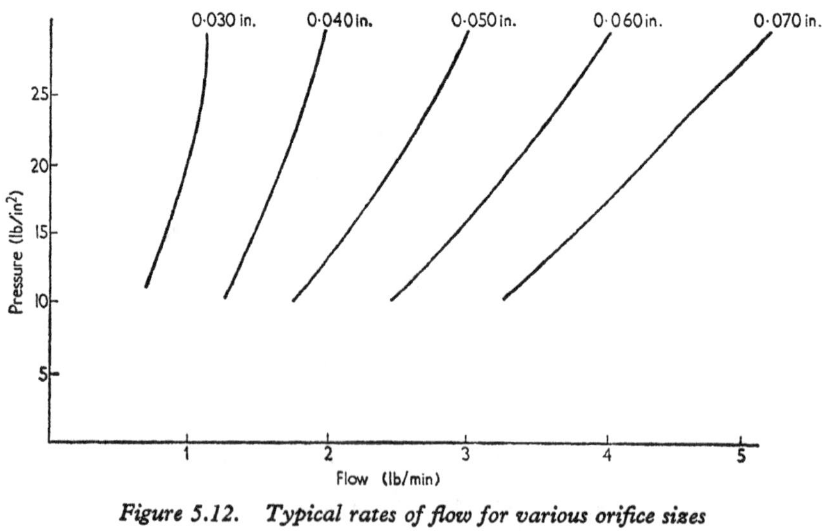

Figure 5.12. Typical rates of flow for various orifice sizes

After the emulsion has passed the metering point it is applied to the jute. Two methods of application are available, sprays or weirs, the former being used with pressure-fed systems drawn from a storage tank and the latter with the ultrasonic unit. Sprays are of the orifice-plate type with either a single central hole or a ring of smaller holes drilled vertically or at an angle at the semi-radius. The quantity of emulsion which will flow through a spray of this type depends upon the pressure at which the emulsion is delivered to it. The higher the pressure the greater is the flow, particularly with sprays with large holes. Figure 5.12 shows the rate of flow for a single-orifice spray operated at different pressures and various orifice diameters and it will be apparent that care must be taken when renewing sprays that the correct size of orifice be fitted.

In the weir method the emulsion trickles down grooves cut in the face of a small metal trough and on to the jute. Whichever method is used it is essential to see that the sliver is covered completely from side to side so that no fibre passes without getting its share of emulsion.

ROLL FORMING

The final demand imposed on the spreader is to provide a sliver in a form suitable for the next stage, carding. The sliver emerges from the nip of the delivery rollers, passes down the conductor, where it is sprayed with the emulsion, then enters the roll-former. The roll-former builds up a close-packed spiral of sliver, hydraulic or air pressure being used to make a dense, compact roll about 4 ft in diameter and 6 in. across the face.

When the roll is of the required size it is ready for doffing. (This is the term used in all textile processes for the action of removing full packages from a machine.) The exact moment of doffing can be decided by the diameter of the roll or by its length. The first method gives rolls of constant weight (or nearly so) whose lengths vary inversely as the count of the sliver; the constant length method gives rolls whose gross weight varies directly as the sliver count. This latter method is useful for routine process checking because if the length is fixed and known then, by weighing the roll, one has a ready means of checking the sliver count.

Depending on the type of roll former in use the rolls may be doffed automatically without the spreader stopping, or it may be necessary to doff the rolls manually, in which case the spreader is stopped.

After the rolls have been doffed they are laid aside for 24–48 hr to mature or condition. The moisture and oil added at the spreader are always, in spite of all the precautions taken, very irregularly distributed, but when the rolls are allowed to stand the moisture and oil become more evenly spread on the fibres. If too much water has been added or if the water is uneven and patchy on the sliver, lapping often occurs at the cards, i.e. the damp fibres stick to the rollers at the feed or delivery of the cards and travel round with them, a wad of fibres builds up, and the machine has to be stopped so that the jute can be cut off.

Spreader rolls may exhibit spontaneous self-heating just as root cuttings do. The benefits accruing (if there are any) have been the subject of debate. It is claimed that heating leads to better carding and cleaner yarn and other desiderata, but in spite of carefully controlled tests and many hours of observation at all stages under practical mill conditions no improvements in the process or the product have been seen when heated and cold jute have been processed. Nevertheless, many manufacturers hold that over long periods there is a definite improvement in processing when the spreader sliver is allowed to heat. The optimum conditions for heating vary from mill to mill but it is known that heating is stimulated by applying hot emulsion, applying sufficient moisture, building a large stack of rolls in a draught-free site, and using a protein-activated emulsifying agent which provides a readily assimilated food supply for the micro-organisms.

The technical details of the spreader vary according to the machinery maker and production requirements, but the figures shown below are typical of spreader operation in the United Kingdom.

Gill width	23½ in.
Pitch of pins:	
Fast chain	1 in.
Slow chain	1⅜ in.
Pin projection	5 in.
Pin diameter	⅜ in.
Feed speed	10–12 ft/min
Delivery speed	200–225 ft/min
Production	1,800–2,400 lb/hr
Range of drafts	6–12
H.P. to drive	12
Sliver roll weight	200–300 lb

Sliver roll length	300–400 yd
Sliver roll formation time	6–8 min
Sliver count	45–60 lb/100 yd (225–300 ktex)

BATCHING FOR SACKING YARNS

These classes of yarn are normally prepared for carding by passing through a softener. The raw material for such yarns is invariably of a lower grade than that required for hessian and similar yarns, that for sacking weft in particular being poor. For sacking warp the low grades of jute in kutcha bales are brought to the spreader feed where selectors make up stricks which are laid on the softener feed. The softener is a long machine comprising 64 or 72 pairs of cast iron fluted rollers, the lower of the pair being driven from a side-shaft and the upper, spring-loaded one by contact with the lower of the pair. Figure 5.13

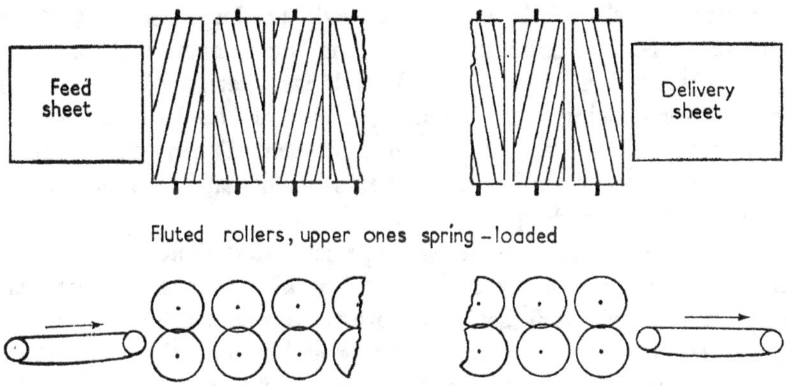

Figure 5.13. Jute softener

shows a diagrammatic view of a softener. The jute is flexed between each pair of rollers and is made softer, some of the loose dust and dirt falls off, and pieces of bark and stick become broken, making their removal at later stages easier. About two-thirds of the way along the rollers the emulsion is dripped on to the jute over a simple gravity-fed weir. As the jute is not fed to this machine in a continuous manner as it is on the spreader, there are gaps in the material and some of the emulsion falls straight through between the rollers; in addition it can drip from the jute itself as it proceeds towards the end of the machine. In order to collect this excess emulsion there is a sump beneath the machine. From here the excess is pumped back to storage through various filters to extract all the dirt and waste which inevitably finds its way into the sump.

Operatives standing at the end of the softener collect the jute as it comes off the machine, give the stricks a half-twist, and place them on a table. On the other side of this table another set of workers cut the root ends of the jute off (it will be remembered that in kutcha assortments the heavy, barky root end is not cut off). The root ends, or cuttings as they are called, are laid aside in special stalls to mature; these will be used later for sacking weft. The long jute is conditioned for 24–48 hr and is then ready for feeding to the breaker cards.

The quantities of cuttings from the warp batches are not sufficient for all the sacking weft yarns, so further supplies of cuttings must be obtained and, these, together with old bale-ropes and any tangled ravelled jute which is unsuitable for higher qualities are put through a softener. This material joins the cuttings from the warp batches in the maturing stalls. If the piles in the stalls are large enough ($1\frac{1}{2}$ tons or more) self-induced heating will occur and temperatures of 60° C may be reached in 9–10 days. This longer period of maturing is required for sacking weft batches because of the large amount of hard, rooty, barky material contained in them. During the maturing period the bacterial activity softens this harsh material, rendering its removal easier at carding.

The bins are usually built with specially slatted floors to allow a gentle circulation of air, a factor which is known to encourage heating. The heat which is generated arises from the growth of micro-organisms left on the fibre after retting and though the exact nature of this activity is not completely understood it is thought that the micro-organisms oxidize some of the natural fats and waxes in the jute, generating heat in the process. Tests have shown that up to 14 days in the pile cause no loss of fibre strength. After their sojourn in the stalls the cuttings are ready for feeding to the weft teaser cards.

Lattice feeders are now available which give an improved method feeding cuttings to a softener, Figure 5.14 showing one such system used by a Douglas Fraser Ltd machine. The cuttings are thrown on to a short conveyor which carries them forward to an inclined spiked lattice. The lattice carries the jute upwards past an evener roller which knocks any excess material back down the lattice. The cuttings continue until they are stripped from the lattice by a rotating bladed doffer and fall in an even stream on to the softener feed sheet. Adjustments to the speeds of the various components can be made and the clearances between the different parts can be altered to provide a range of operating conditions.

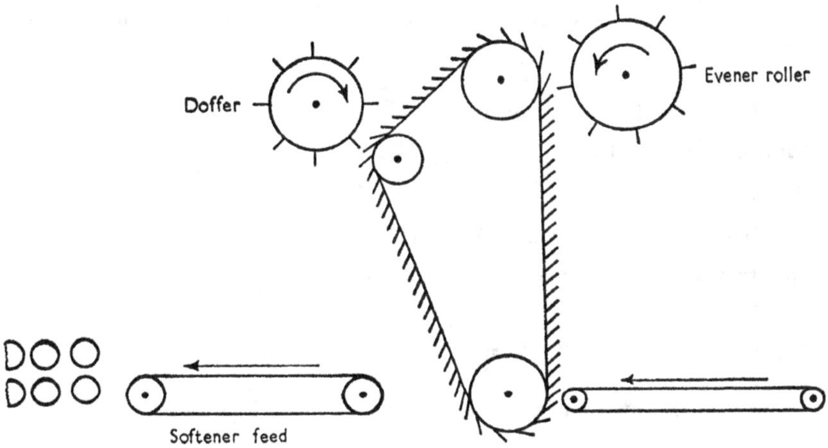

Figure 5.14. Root cuttings feeder

James Mackie and Sons Ltd produce a type of softener for use in conjunction with a lattice feeder similar to the one just described. It differs from the traditional machine in that it has two sets of 24 rollers and, between each set, a cuttings opener, consisting of a drum with coarse pins on its surface. The cuttings come along between the nips of the first set of rollers to meet the opener whose pins effect some degree of opening and begin the work of breaking down the hard, barky material. The opener then passes the cuttings to the next set of rollers where further flexing and softening occurs. The emulsion is added in two stages—the first application being just as the material enters the second set of rollers immediately after the cuttings opener. This dual application is said to lead to a more even distribution of moisture and oil on the jute. After the jute leaves the nip of the last pair of rollers it falls on to another conveyor which carries it to the maturing bins where it lies for a period so that the bacterial activity can soften the rooty material in the usual manner. It should be noted that this system requires the minimum amount of manpower, conveyors being used wherever possible. The remainder of the machines in this special range will be dealt with later.

SPREADER CALCULATIONS

The following worked examples are typical of those met with in working with a spreader or softener.

What is the delivered count of the sliver under the following conditions?

Raw jute feeding rate (lb/min)	=	27
Emulsion flow (gal/hr)	=	32
S.G. of emulsion	=	0·97
Length of sliver in a roll (yd)	=	450
Time to form a roll (min)	=	7·2

$$\text{Emulsion flow} = \frac{32\times 9{\cdot}7}{60} = 5{\cdot}2 \text{ lb/min}$$

$$\text{Total delivery} = 27{\cdot}0+5{\cdot}2 = 32{\cdot}2 \text{ lb/min}$$

$$\text{Delivery speed} = \frac{450}{7{\cdot}2} = 62{\cdot}5 \text{ yd/min}$$

$$\text{Sliver count} = \frac{32{\cdot}2\times 100}{62{\cdot}5} = 51{\cdot}5 \text{ lb/100 yd}$$

If leaders were used how many leader rolls per hour would be needed if the draft is 12?

$$\text{Feed speed} = \frac{62{\cdot}5}{12} = 5{\cdot}21 \text{ yd/min}$$

$$\text{Length on a roll} = 450 \text{ yd}$$

Therefore

$$1 \text{ roll will last } \frac{450}{5{\cdot}21} = 86{\cdot}5 \text{ min}$$

But two leaders are always required, so

$$\text{Rolls per running hour} = \frac{2\times 60}{86{\cdot}5} = 1{\cdot}39$$

What is the moisture regain of the sliver produced when the raw jute regain is 16 per cent and the application rate is 22 per cent, the emulsion being a 32/68 mix? At this regain the sliver count is

320 ktex, what will it be if the raw jute regain drops to 13 per cent? Raw jute fed consists of

Fibre	100 parts
Moisture	16 parts
Total	116 parts

Amount of emulsion added, 22 per cent of 116 = 25·5 parts

Of this quantity, 68 per cent is water and therefore the amount of water added to the jute is 17·3 parts (68 per cent of 25·5) and so the delivered sliver consists of

Fibre	100 parts	Raw jute
Moisture	16 parts	
Water	17·3 parts	Emulsion
Oil	8·2 parts	
Total	141·5 parts	

Hence

$$\text{Sliver regain} = \frac{16+17\cdot3}{100}\times 100 = 33\cdot3 \text{ per cent}$$

If the raw jute regain falls to 13 per cent then the sliver regain will become 30·3 per cent (13+17·3) and the sliver count will drop to

$$\frac{320\times 138\cdot5}{141\cdot5} = 318 \text{ ktex}$$

The following information has been collected during a test:

Weight of jute on barrow at start of test = 1,415 lb
Weight of jute on barrow at end of test = 632 lb
Roll former speed = 63 yd/min

Roll weights	*Roll formation time*	
260 lb	6 min	50 sec
255	6	22
268	7	2
261	6	30

Average tare of metal roll-centres = 24 lb
Draft constant = 360
Draft pinion = 32 teeth

Find (1) the sliver count,
(2) the emulsion flow rate,
(3) the emulsion application rate,
(4) the weight per unit length on the feed sheet.

$$\text{Average roll weight} = \frac{260+255+268+261-4\times24}{4} = 237 \text{ lb}$$

$$\text{Average roll formation time} = \frac{6{\cdot}8+6{\cdot}4+7{\cdot}0+6{\cdot}5}{4} = 6{\cdot}68 \text{ min}$$

$$\text{Length on roll} = 6{\cdot}68\times63 = 420{\cdot}8 \text{ yd}$$

$$\text{Therefore, Sliver count} = \frac{237\times100}{420{\cdot}8} = 56{\cdot}5 \text{ lb/100 yd (283 ktex)}$$

$$\text{Weight of raw jute fed} = 1{,}415-632 = 783 \text{ lb}$$

$$\text{Total feeding time} = 26{\cdot}73 \text{ min } (4\times6{\cdot}68)$$

$$\text{Feeding rate} = \frac{783}{26{\cdot}73} = 29{\cdot}3 \text{ lb/min}$$

$$\text{Delivery rate} = \frac{237}{6{\cdot}68} = 35{\cdot}4 \text{ lb/min}$$

$$\text{Therefore, Emulsion flow rate} = 35{\cdot}4-29{\cdot}3 = 6{\cdot}1 \text{ lb/min}$$

$$\text{Application} = \frac{6{\cdot}1\times100}{29{\cdot}3} = 20{\cdot}8 \text{ per cent}$$

$$\text{Machine draft} = \frac{360}{32} = 11{\cdot}3$$

$$\text{Feed speed} = \frac{63}{11{\cdot}3} = 5{\cdot}58 \text{ yd/min}$$

$$\text{Weight per unit length on the feed sheet} = \frac{29{\cdot}3}{5{\cdot}58} = 5{\cdot}25 \text{ lb/yd (2,625 ktex)}$$

CHAPTER SIX

Carding

THE primary function of the cards is to convert the reeds of jute into a uniform supply of fibrous material which can then be drafted and finally twisted into yarn. It is perhaps at the cards that the most dramatic change in the appearance of the jute is seen; when it is passing into the breaker card the reeds from the stems of the plants can easily be identified and the whole feed is coarse and uneven, but by the time the jute has passed through the breaker and finisher cards it has been transformed into a thin web of separate fibres emerging as a fleece which is then condensed into a sliver. Besides this essential task, the cards begin the work of weight reduction by drafting and weight levelling by doubling.

Like the batching process, the carding system in use depends upon the class of fibre being worked and, in general, two methods are adopted, one for long jute and one for short material. In practice this reduces to one method for hessian and sacking warp yarns and another for sacking weft yarns, the former comprising two carding passages and the latter three. Jute and its allied fibres are invariably carded on roller and clearer type cards based on those used in the flax trade. The heart of the machine is a large cylinder 4–5 ft in diameter and covered with small pins set at an angle to its surface. Arranged round the periphery of this main cylinder are complementary pairs of smaller rollers clad also with pins, these rollers being known as the workers and strippers. The pins of the worker are set to work against those of the main cylinder whereas the stripper pins are set in the same direction as the cylinder pins (see Figure 6.1). As the cylinder, workers, and strippers rotate, their pins split and open the jute which is passing between them. On a breaker card there are usually only two pairs of workers and strippers but finisher cards commonly are made with four or five pairs. Another pinned roller is required at the delivery of the machine to pluck the carded fibre from the pins of the main cylinder and pass it to the delivery rollers, this roller being known as the doffer from its function of doffing the fibre from the cylinder.

Jute cards are classified according to the direction in which the cylinder pins are travelling when they meet the jute for the first time

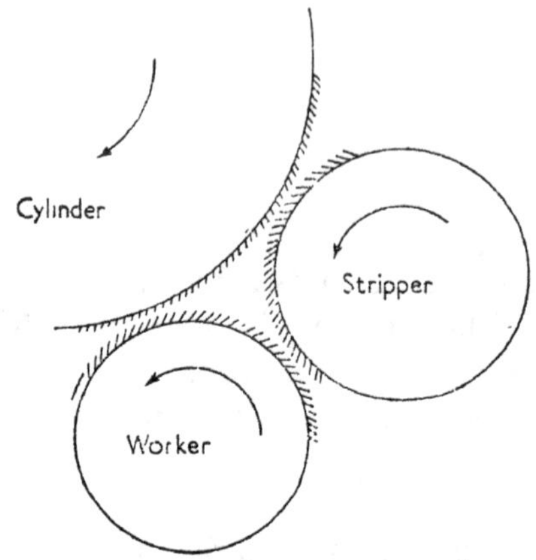

Figure 6.1. Worker/stripper/cylinder pin action

at the feed side of the machine and according to the amount of the main cylinder circumference which is utilized.

Upstriking, i.e. the pins of the cylinder approach the feed from underneath and strike up into the jute.

Downstriking, i.e. the cylinder pins approach the feed from the top and strike down into the fibre.

Half-circular, i.e. the jute travels half-way round the cylinder in its journey from the feed to the delivery and thus the feed and delivery are approximately 180 degrees apart.

Full-circular, i.e. the feed and delivery are almost side by side and the jute travels through nearly 360 degrees inside the machine.

Breaker cards are commonly downstriking and half-circular, and finishers, downstriking and full-circular. Upstrikers are mainly used to card low-grade material of short length for if this kind of fibre is handled on a downstriker card there tends to be a large amount of short fibre dropped beneath the machine. If the card is upstriking this material is held in with the bulk of the jute. With this type of card the sliver tends to be specky and dirty because of the accumulation of short fibre plus all the small pieces of bark and stick which would otherwise have fallen underneath the machine.

CARDING SYSTEM

Hessian yarns are given two carding passages, the breaker card handling 500–800 lb/hr and the finisher 350–450 lb/hr. The breaker card is fed from spreader rolls and the finisher card from breaker rolls. Between the two machines there is an effective draft of 3–4 to reduce the heavy spreader sliver to a count suitable for the first drawing stage.

Sacking warp material is given two carding passages over breakers and finishers which are similar to hessian-type cards, but the method of feeding the breaker differs. The jute for this grade of yarn is usually passed over the softener and it is converted to a continuous sliver at the breaker card instead of at the preceding stage as in the case of the hessian qualities. Just as the spreader is fed by a dollop of a certain weight to a pre-determined clock length, so the breaker is dollop-fed by hand. The breaker clock length, however, is only about 12 or 15 yd and the dollop weighs about 35 lb. The jute is taken from the conditioning site to the breaker cards where dollop-weighers make up bundles of fibre, equal in weight to half the dollop, which are then placed evenly on the card feed sheet by the breaker feeder at such a rate that the half-dollop is fed in half of one revolution of the gear-driven pointer at the top of the feed sheet. If, for example the clock length is 11·5 yd, the dollop 38 lb, and the card draft 17, then the delivered sliver count will be

$$\frac{38 \times 100}{11{\cdot}5 \times 17} = 19{\cdot}5 \text{ lb/100 yd}$$

When sacking weft is being prepared an extra carding machine is used to convert the bale ropes, yarn, and cloth waste used for this quality into a fleece or tow before mixing it with root cuttings and some long jute at the sacking weft breaker card. This machine is known as a teaser or waste breaker card and consists of a cylinder, two worker/stripper pairs, and a doffer, all clad with strong, coarsely set pins. The waste material is fed by hand to the teaser and the delivered fleece is allowed to fall to the floor or into bags. At the sacking weft breaker card, this tow is weighed out in the required proportions and mixed with the root cuttings and low marks of long jute which form the rest of the batch for this yarn. From the breakers the sliver passes to the sacking weft finishers which again are more sturdily pinned than hessian finishers and can be more heavily loaded, having a production rate of about 700 lb/hr.

THE JUTE BREAKER CARD

The breaker card is a particularly important machine in the jute processing system for it is here that the very basis of yarn quality, the average fibre length and fineness, is determined. Breakers are usually half-circular and downstriking and have two pairs of workers and strippers, Figure 6.2 showing a sketch of one such machine.

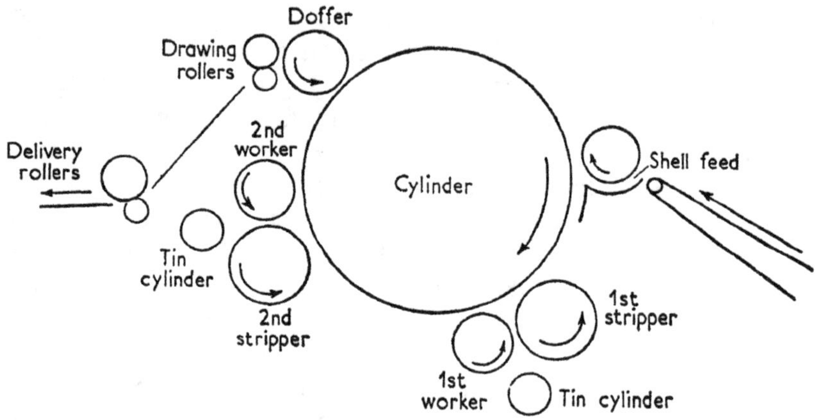

Figure 6.2. Two-pair downstriker breaker card

The carding action of converting the reeds into a fibrous fleece is not easy to examine since the machine must be enclosed with shrouding while it is running but, nevertheless, a full understanding of the process is essential. The description that follows refers particularly to a hessian breaker card but the same principles apply for sacking cards.

The rolls of spreader sliver, 6–8 in number, are fed on to the feed sheet from a creel at floor level and the material passes up towards the feed rollers of the breaker. The jute then enters the machine through what is known as a 'shell' feed. This consists of a pinned feed roller with backward-facing pins and a cast iron shell which is shaped to follow the circumference of both the feed roller and the main cylinder and forms a sharp edge between the two curvatures over which the jute must pass. The jute enters the space between the pinned feed roller and the shell and travels towards the swiftly moving pins of the main cylinder. When the leading ends of the reeds meet these fast-moving pins (which strike down into the jute because of the angle at which they are set in the cylinder and the direction of cylinder rotation) they are split, opened out, and converted into a fibrous 'beard', which hangs down between the lower part of the shell and the cylinder. More jute is fed forward and the beard becomes longer and the reeds are opened progressively. The longer the jute which is fed into the breaker the longer is this beard and to accommodate long reeds the first stripper is set 4 or 5 ft away from the shell; if this were not done there would be a danger that the reeds would become trapped between the pins of the 1st stripper and the cylinder and be pulled rapidly into the card without being sufficiently opened. While this combing, splitting, and opening is going on at the shell feed, the backward-facing pins of the feed roller exert a grip on the reeds and, ideally, each part should receive the same carding action. Unfortunately this does not always happen in practice. The feed roller pins can only exert a restraining influence on the jute provided there is enough bulk of material between them and the shell and if the incoming supply of fibre is not maintained then the jute which is between the roller and the shell will be pulled rapidly into the card over the shell edge and will not receive its fair share of carding—this action is known as 'gulping' the sliver. On breakers fed from spreader sliver the material normally enters the machine crop end first and if a thin section of spreader sliver comes along then gulping may occur and the root end will be dragged past the shell. Since it is important that this part of the reed should be carded properly, this constitutes a source of poor quality. If, on the other hand, the card is being fed by hand on the dollop system, as for sacking grades, the stricks are fed up the feed sheet root end first and this part will almost always receive the necessary degree of carding at the shell. When gulping occurs it is the lighter crop end which misses the combing action at the shell.

The clearances between the shell and the feed roller, the shell and the cylinder, and the feed roller and the cylinder can be altered to give a selection of operating conditions. These settings can have a considerable bearing upon the average fibre length in the card sliver and since it is axiomatic that as long a fibre length as possible is desired, the shell settings assume considerable importance. In general, any adjustment which exposes the jute to prolonged combing at the shell will reduce the fibre length, though naturally sufficient time must elapse to allow the essential opening, splitting, and cleaning to be achieved. Alterations in the combing time come about through changes in the rate of fibre feed. For example, if the fibre is fed at 8 ft/min, then each inch of the reed will pass over the shell edge in $\frac{1}{96}$ min, but if the feed rate is altered to 16 ft/min, then each inch will pass over the shell in $\frac{1}{192}$ min, i.e. half the time. Like the spreader, the breaker card is a constant delivery speed machine and when the draft is changed it is the feed speed that is altered, thus changes in the length of time the jute is exposed to combing will arise from changes in the card draft. Table 6.1 shows the results of tests made to find the effect of different breaker card drafts on the average fibre length.

TABLE 6.1. EFFECT OF BREAKER DRAFT ON FIBRE LENGTH

Draft	*Feed speed (ft/min)*	*Average fibre length (in.)*	
		Jute A	*Jute B*
24	8·1	2·4	2·7
18	11·2	2·6	2·7
12	16·3	2·7	2·9

Clearly, the greater the draft the more fibre breakage takes place at the shell and as the quality of the yarn depends *critically* upon fibre length it might be thought that a substantial improvement in the process would be obtained by using low breaker drafts. Unfortunately, undesirable side-effects come into operation when the breaker draft is low; for instance, if the same count of spreader sliver is used then a heavy breaker card sliver is made and this throws a heavy load on the finisher card and the drawing frames. Alternatively, the same count of breaker sliver could be made but this would be accompanied by a drop in production and more breaker cards would be required to put through the same tonnage of jute. As in so many cases, a compromise must be reached between acceptable quality levels and the demands of production.

The other important variable is the clearance between the shell and the feeder roller. The greater the distance between the two the lighter is the grip on the jute and the more easily is it dragged into the card by the cylinder pins. This has the same effect as reducing the time of combing, i.e. less breakage occurs. Table 6.2 gives a summary of the results of an experiment where two shell/feed-roller settings were used and the yarn properties examined.

The jute processed with the closer shell setting had a shorter average fibre length and poorer yarn properties. However, it cannot be deduced from this that the wider the feed/shell setting the better will be the quality of the yarn since there must be sufficient restraining

TABLE 6.2. EFFECT OF SHELL SETTING ON YARN QUALITY

Yarn properties	*Shell/feed-roller clearance* $\frac{3}{8}$ in.	$\frac{1}{2}$ in.
Count (lb/sp.)	7·9	7·9
Breaking load (lb)	7·2	7·5
Coefficient of variation of breaking load (per cent)	18·3	17·9
Calculated minimum strength (lb)	3·2	3·5

action at the feed to hold back the jute and allow enough time for the cylinder pins to do their work.

While dealing with the shell feed it is convenient to consider the effect of the physical properties of the jute on the fibre length obtained. It has been said already that the cylinder pins strike strongly into the projecting reeds at the edge of the shell and in so doing comb, split, and open the fibre complexes into more or less single units. As may be imagined the force exerted on the jute to achieve this function is considerable and therefore any fibre which is weak and 'brittle' would not be able to withstand this rigorous treatment as much as a strong fibre with some 'give' in it. These terms 'brittle' and 'give' are more correctly replaced by 'inextensible' and 'more extensible'; that is to say, a fibre which cannot be extended much without breaking has a low resistance to sudden impulsive loads and another which can extend more before it breaks will be better able to withstand such forces. Thus it is this property of extensibility which is important in a fibre in regard to its ability to be carded into a fibrous mass while retaining a good average fibre length. It will be recalled that, when the grading of raw jute was dealt with, one of the factors characterizing a good quality jute was good elasticity; now it can be seen that this has a very important bearing on the standard of the yarn which can be spun from it. Table 6.1 shows two fibre lengths obtained from the same conditions on the breaker card from two grades of jute, the better grade, B, having a higher extensibility, suffered less fibre breakage than the poorer grade.

In the early attempts to spin jute at the beginning of the nineteenth century it was only the line fibre which could be hand-spun into yarns at all—line fibre being the longer fibre extracted by a combing process from the bulk of fibre. With jute this meant that there was only a small quantity of line usable, leaving large amounts of tow containing very short fibres which could not be worked into yarn. One or two trials were undertaken in which jute was passed over a flax card but the average fibre length was so low that nothing could be done with the material and, besides, the dust and waste was excessive. It was not until batching was found to have a beneficial effect that the commercial exploitation of jute could begin. Unknowingly, the pioneers of jute spinning had conferred upon the fibre that very property which permits the mechanical carding of jute to be carried out successfully—increased extensibility. Batched jute has a higher extension at break than unbatched jute and, as a result, withstands the fierce action at the shell feed better and excessive fibre breakage is avoided. The difference in fibre length obtained with batched and unbatched jute is small but even a change of $\frac{1}{4}$ in. in the average can have an extremely large effect in the yarn, a lesser fibre length increasing the variability markedly. Since the working strength of a yarn is the strength of its weakest point, this results in a much inferior yarn.

From the preceding paragraphs it will be apparent that the carding action at the shell feed is one of great importance as it is here that the average fibre length is largely decided, but despite the robust operation at the shell all the reeds are not split and opened by the time the jute has passed completely into the card and further carding is required; this is carried out by the combination of the workers and strippers. The essential feature of the carding operation at the workers and strippers is one of combing, teasing, and splitting as the fibre is transferred first from the cylinder to the worker, then from the worker to the stripper and finally from the stripper back on to the cylinder. As the pins of the stripper point in the same direction as those on the cylinder the bulk of the jute descending on the cylinder passes between the cylinder and the stripper though there is a certain amount of build-up where the two rollers meet. The jute then comes in contact with the backward-facing pins of the worker and as the surface speed of this roller is very low compared with that of the cylinder (30 ft/min compared with 2,500 ft/min) the fibres are arrested by the worker pins. Those fibres which are firmly held by the cylinder pins pass underneath the worker and continue until they meet the next worker/stripper pair, but the remainder are pulled away from the cylinder pins by the worker pins. There is no clear-cut transition from one set of pins to the other but rather an indefinite fleecy mass forms in which the longer fibres are tugged between and round pins, and the shorter fibres dragged through longer ones still anchored in the cylinder pins. This fibre entanglement and inter-fibre movement continues the splitting and opening work begun at the shell and, in addition, much of the hard, barky particles adhering to the fibres are knocked off. The fibres do not lie as a uniform fleece on the pins of the worker, some being held by the pins but the greater majority resting quite lightly on the top of the pins. It will be appreciated that inside the shrouding of the card air currents and eddies are set up by the rapidly rotating cylinder; these tend to blow the fibres into a fleecy conglomeration on the workers and strippers. This fibre mass continues round the worker until it comes under the influence of the stripper where the pins, by virtue of their direction and greater surface speed, lift the upper layers of the fibre mass off the worker pins, splitting and opening it as they do so. The more firmly held fibres pass round the workers again until they meet the faster-moving cylinder pins and the cycle is repeated. Once the bulk of the fibre has been transferred from the worker to the stripper it continues round on the stripper until it meets the cylinder where it is removed from the stripper by the cylinder pins working back-to-back with the stripper pins. In this way the fibre networks are gradually split into a fibrous mass and, in addition, there is some degree of mixing inside the card as some of the material is held back and deposited on top of fresh. Indeed, it is possible for a bunch of fibres to travel several times round the worker/stripper pair before it passes on with the cylinder.

After leaving the second worker/stripper pair the jute meets the doffer whose function is to strip the jute off the cylinder and pass it to the nip of the drawing rollers and so out of the machine. Its action is similar to that of a worker, its pins plucking the fibre away from the cylinder surface. Ideally, the doffer should remove all the jute from the main cylinder so that the latter can approach the shell feed with clean pins ready for the maximum amount of carding but this seldom happens and some fibres evade the doffer's action and travel round with the cylinder. Most of the jute, however, is caught by the doffer and moves round with it until it meets the drawing rollers. These normally have about twice the linear speed of the doffer. The fibre is caught as a thin tenuous web in the nip of the drawing rollers and the doffer pins, by virtue of their downward movement relative to the drawing rollers, give up the fibre smoothly. The web emerges from the drawing nip and passes down a V- or U-shaped sheetmetal condenser at the foot of which is an opening which the jute passes through to form a sliver. Once through this hole it passes between two delivery rollers, the top one of the pair being heavy enough to compress the jute into a compact sliver.

Mention must be made of two other rollers found on the breaker card—the tin rollers. These are light hollow rollers situated at each worker/stripper pair (see Figure 6.2) for the purpose of reducing the amount of fall-out below the card. As the fibre fleece is travelling round with the workers and strippers it stands proud from the surface and there is a tendency for the long fibres to fall out of this fibrous mass. The tin rollers prevent this by containing the mass without interfering with the essential operations in any way.

Facilities are available to alter the clearances between the workers, strippers, and cylinder pins and to change the speed of the workers and strippers relative to the cylinder, but these settings or speeds do not appear to be critical. In tests where the worker speed was changed from 27 ft/min, to 111 ft/min, there was no alteration in the average fibre length or the physical properties of the yarn. This is perhaps not surprising, for it is not the roller's absolute speed that matters but rather its speed relative to that of the cylinder. When the cylinder has a linear speed of 2,500 ft/min and the worker's speed is 27 ft/min, the relative speed of the two is $2{,}500-27=2{,}473$ ft/min; increasing the worker surface speed to 111 ft/min only alters the relative speed to 2,389 ft/min, i.e. increasing the speed of the worker by a factor of around 4 changes the relative carding speed by only 3·4 per cent.

In the preceding paragraphs the impression may have been given that the pins of the rollers are carrying large quantities of fibre but this is not the case. It is interesting to calculate the weight of fibre on each square foot of the rollers of a breaker card, taking an average throughput of 12 lb/min, average roller surface speeds, and assuming that the cylinder carries fibre only over two-thirds of its surface, the feed roller carries fibre over one-third of its surface, and the doffer, over two-thirds. (These assumptions lead to the highest density of fibre on the rollers being calculated.)

Roller	*Loading (lb/ft^2)*
Feed	0·671
Cylinder	0·001
Workers	0·050
Strippers	0·005
Doffer	0·022

The figures illustrate how the jute is transformed from a thick heavy layer of reeds at the feed, travelling at perhaps 9 ft/min, to a thin web on the cylinder, travelling at about 40 m.p.h., and give one some indication of the forces applied to the fibres.

THE FINISHER CARD

Figure 6.3 shows the layout of a finisher card suitable for hessian, carpet, and sacking warp yarns. The machine is designated 4½ pair, from the number of pairs of workers and strippers; full circular, from the fact that most of the cylinder surface is usefully employed in carding; double doffer, from the two doffers; and down-striking.

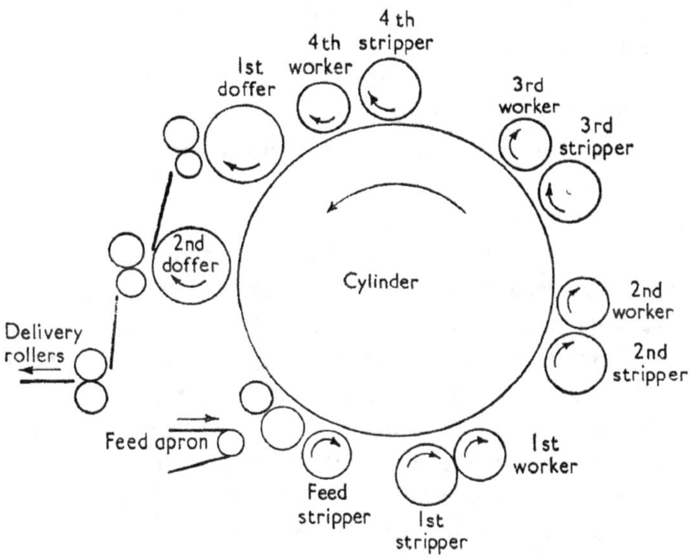

Figure 6.3. 4½-pair full-circular finisher card

The rollers and the cylinder are pinned in the same manner as the breaker card, but because the jute is in a more open state by the time it reaches the finisher the pins are somewhat finer and set closer together. The commonest type of feed arrangement is the 'pinned plain', i.e. the feed roller is clothed with pins but the roller immediately above it is not. The pinned plain feed is met with universally on hessian and carpet quality cards and very often on sacking warp cards. Two other types of feed are met with, the 'shell' feed and the 'double-pinned' feed. The shell feed is similar to that on the breaker with the shell set about $\frac{1}{4}$ in. from the feed roller; this feed is associated with heavy card loadings and is usually confined to sacking weft cards. In the double-pinned feed the sliver passes between two pinned rollers; with this method only light card loadings are possible.

The action of the workers and strippers in the finisher is the same as in the breaker, and the finisher, therefore, continues the work of weight reduction by drafting, reducing weight irregularities by doubling, splitting the fibre networks, and cleaning.

When good quality fibre is being worked most of the fibre complexes are opened sufficiently at the breaker and the finisher's usefulness lies in drafting and doubling; but when low grade, dirty jute is being carded the finisher does much to open the material and remove bark and stick.

The settings of the various rollers can be varied and the speeds altered by changing speed pinions and pulleys in the drive. The clearances and speeds do not appear to be critical since tests where the worker speed was altered progressively from 30 to 97 ft/min for poor and good jute showed that there was no effect on either the spinning performance or the yarn quality, and others where the roller settings were changed by factors of almost 3 indicated that there was only a slight advantage to be gained from using the maker's recommended settings. If, however, the card is overloaded, there is a deterioration in sliver quality as shown by poorer spinning performance and low yarn quality. Too great a load in the card forces the fibres on to the pins which cannot then exercise their proper functions and the essential splitting of the fibre complexes ceases and long aggregations of fibre pass forward into the drawing stages and yarn. These may be seen at the finisher card delivery when the fleece is passing down the conductor which, incidentally, is a good place to examine the fibre from the point of view of the effectiveness of carding.

FINISHER CARD DRAWING-HEADS

At the delivery of a finisher card there may be a short drawing-head attachment comprising a pair of feed rollers, a short lattice of pinned bars and a pair of delivery rollers. The sliver enters the feed roller nip and then passes forward on to the moving pinned sheet. The pins are intended to pierce the sliver and control fibre movement as the material approaches the delivery rollers. On the drawing-head there is a draft of about 2 and the head functions mainly as count reducer for systems having only two drawing passages. In addition, the drawing head straightens the fibres somewhat and orients them parallel to the axis of the sliver. The major disadvantages in drawing-heads are associated with the speed at which they must run. Since high speed is necessary, the pinned lattice must be of the push-bar variety and even this must be run at such a speed that mechanical wear and tear is high and the pins cannot pierce the sliver correctly so that very often it rides on top of the pins instead of within them and uncontrolled fibre movement occurs.

CARD PINNING

The card pins are set in staves or lags which are screwed to the rollers, the staves being curved to match the rollers and extending across the roller face. The pins are set at an angle to the surface of the staves and are arranged in rows both horizontally and circumferentially, the latter rows being staggered by half the pin pitch so that when the surface of the roller is examined the pins appear in a diamond formation. For hessian cards the pins are set in beech-wood staves but for cards designed to handle harsh material the staves are metal.

The density of pinning becomes greater as the material being carded becomes finer. That is to say, finisher card pinning is finer than breaker card pinning, hessian pinning finer than sacking, and so on.

To maintain efficient carding the pins must be kept sharp. The points become blunted with use and grooves develop in the sides of the pins as a result of the fibres being pulled round them; collections of fibre may gather in these grooves to be released as small tight balls or 'neps' which persist through to the yarn. A regular scheme of pin renewal is required, the frequency depending on the class of fibre being worked, harsh fibre being particularly hard on the pins. It is customary to renew half the cylinder staves at a time and some indication of the intervals at which this should be done is given below.

Hessian and sacking warp grades of fibre

Breakers	800 hr
Finishers	2,500 hr

Sacking weft grades of fibre

Breakers (waste)	400 hr
Breakers	700 hr
Finishers	2,000 hr

Worker pins should last for roughly 2 years and stripper pins about 3 years on two-shift working.

TABLE 6.3. TYPICAL CARD PINNINGS

Roller	*Pins per square inch*					*Pin angles (degrees)*
	Breakers			*Finishers*		
	Waste	*Sack. weft*	*Sack. warp & hessian*	*Sack. weft*	*Sack. warp & hessian*	
Cylinder	2·5	2·5	3	7	10	70–75
Feed	3	4	4	6	9	50–60
Workers	4	4	5	5	10–16†	30–35
Strippers	4	5	5	5	8–16†	35–40
Doffers	4	5	6	6	12–18†	30–40

† Pinning becomes progressively finer as the delivery is approached.

BREAKER AND FINISHER CARD OPERATING DATA

While individual machinery makers and manufacturers vary slightly among themselves in technical detail, Tables 6.4 and 6.5 represent typical conditions.

TABLE 6.4. CARD SPEEDS

	Breaker	*Finisher*
Cylinder	2,400–2,700 ft/min	2,400–2,800 ft/min
Feed roller	9–14	10–15
Workers	35–50	30–40
Strippers	300–500	300–500
Doffer	75–95	75–100
Drawing Rollers	150–200	150–200

TABLE 6.5. DRAFTS, DOUBLINGS, AND SLIVER COUNTS AT THE CARDS

	Drafts	*Doublings*	*Counts*	
			lb/ 100yd	*ktex*
Hessian breaker	10–20	6–8	18–20	90–100
finisher	10–15	10–12	13–18	65–90
Sacking breaker	10–18	—	18–26	90–130
finisher	12–18	8–10	14–18	70–90

SLIVER PACKAGES

Breaker and finisher card sliver is usually delivered in rolls though certain systems still use cans at the finisher delivery. Roll-forming seldom gives much trouble, the main faults being extremely soft or extremely hard rolls which will not unwind properly at the next stage in the process.

Soft rolls may be due to insufficient pressure on the roll-former, dry sliver, or a low lead between the delivery roller of the card and the roll-former itself. Hard rolls, conversely, arise from too high a pressure on the roll-former, a high lead, and high moisture in the sliver (a defect accompanying the last-mentioned factor would probably be lapping at the drawing roller, i.e. sliver wrapping itself round the roller instead of passing cleanly through the nip).

The roll-former lead over the delivery of the cards varies with the position of the roll-forming mechanism. For convenience in the layout of the carding machines in the space that is available it is sometimes necessary to lead the sliver from the card at the side, front, or rear. This can be done simply by passing the sliver round guides, plates, etc., and, in the rear delivery case, carrying the sliver to the back of the machine by a short conveyor belt acting in conjunction with a guide plate. At all times it is necessary to have the sliver under a slight tension so that there is no danger of the sliver going slack and fouling the delivery. The lead required is of the order

3 per cent for breaker cards.
5 per cent for finishers (front delivery).
13 per cent for finishers (rear delivery).

SPECIAL CARDING SYSTEMS FOR ROOT CUTTINGS AND SIMILAR MATERIAL

In recent years two special ranges of cards have been developed for handling root cuttings and similar material with a view to giving the fibre a thorough carding and distributing it as evenly as possible among better class jute so that the cheaper grade material can be used successfully as a diluent and permit a reduction in the cost of the batch to be made.

(1) *The Fraser system.* This range of machinery consists of two cards, specially designed to suit the nature of the raw material, and ancillary equipment for blending and mixing. The cuttings are taken from their maturing stalls and dollop fed to a lattice feeder similar to that used for feeding the softener (see Figure 5.13). The lattice feeder distributes the cuttings evenly on the feed sheet of the first card, the J1, which acts as a breaker for the cuttings. When cuttings are being processed the usual shell feed does not exercise sufficient control over the short fibre and it tends to gulp into the card instead of being restrained to allow the cylinder pins to do their work; this results in incomplete opening of the cuttings and an inferior sliver, full of hard bark and root. The J1, however, has a large-diameter pinned roller and a concentric feed-control plate which permits a more positive grip to be imposed on the cuttings, ensuring that no uncarded material passes forward. The cuttings are carded and delivered as rolls. The J1 card has a capacity of around 600 lb/hr.

Following the J1 card is an intermediate card, called the J3, which is fed with several J1 rolls and acts as a mixer for the cuttings besides continuing the work of splitting, opening, and cleaning. Like the J1, it delivers its sliver in rolls, but at a slightly lower rate—about 500 lb/hr.

It is at the next stage that the cuttings are intimately blended with long jute. This is done by an attachment to an ordinary shell-feed breaker card called the Sliver Dispersal Unit (S.D.U.) (Plate II). The S.D.U. is situated at the side of the breaker feed sheet with two rolls of J3 sliver sitting in its creel. The cuttings slivers enter the S.D.U. where they meet a drum carrying a series of beater bars which act simultaneously as choppers and blowers by tearing sections of sliver off the roll as it is being fed and blowing the fibres up a chute, mixing the jute into an expanded fibrous mass as they do so. The chute leads to the top of the breaker card feed sheet where an oscillating blade distributes the fibre uniformly on top of the long jute. From the breaker the material progresses to a finisher card in the usual manner, thus the cuttings receive a total of four cardings in their passage from the maturing stalls to the drawing stages.

The J1 card feeder distributes the cuttings evenly over the feed sheet allowing maximum opening and splitting to be effected in the card. This even feed rate allows the feed to be fast and permits the use of finer pinning than usual. It has been found that dense, short pinning on all the rollers may be used and remain clean despite the nature of the raw material.

TABLE 6.6. PIN DENSITIES

	Pins per square in.	
	J1	*J3*
Cylinder	8	10
Feed roller	5	6·4
Strippers	5	8
Workers	5	8
Doffer	5	5

The speeds of the rollers are much the same as those of hessian cards with the exception of the workers which are rather faster and the J1 feed which runs at about 50 ft/min.

A typical system employs three J1 cards and four J3 cards to produce 2,000 lb of material per hour. The J1 has a feed sliver weight of about 100 lb/100 yd and operates at a draft of 4. The J3 is fed by eleven doublings of J1 sliver and produces sliver of about 17 lb/100 yd.

(2) *The Mackie system.* Once the cuttings have matured they are placed in a hopper feeder attached to the 1st (or teaser) card. The hopper bin can hold 250 lb of root cuttings or mixtures of cuttings, bale ropes, and mill wastes; from the bin the fibre passes on to a moving lattice equipped with rotary beaters to knock off excess. The lattice carries it to an electrical weighing point where the material is weighed automatically and dumped on a second lattice at a rate adjusted to the card input speed to give the desired count of sliver at the delivery. This second lattice spreads the fibre on the feed sheet in such a manner that a uniform distribution of jute is presented to the

card pins. The card itself is a 2½ pair machine clad with coarse pins and functions as a sliver former and initiates the break-down of the hard, rooty material. The card delivers into rolls ready for the 2nd (or intermediate) card.

The 2nd card is a 5½ pair, pinned plain feed machine which is fed with 11 rolls of 1st card sliver and has a capacity of 450 lb/hr. The material is given a further opening and cleaning treatment in this card and it emerges from it in a suitable form for blending with long jute or passing directly to the drawing stages for sacking weft. If it is to be blended with long jute the sliver rolls are placed in a special dual creel which has positions for the cuttings rolls and the long jute rolls (Plate III). Blending may be carried out at either the hessian breaker card or the finisher card; if at the latter the poorer degree of blending must be accepted. The cuttings sliver is presented to the

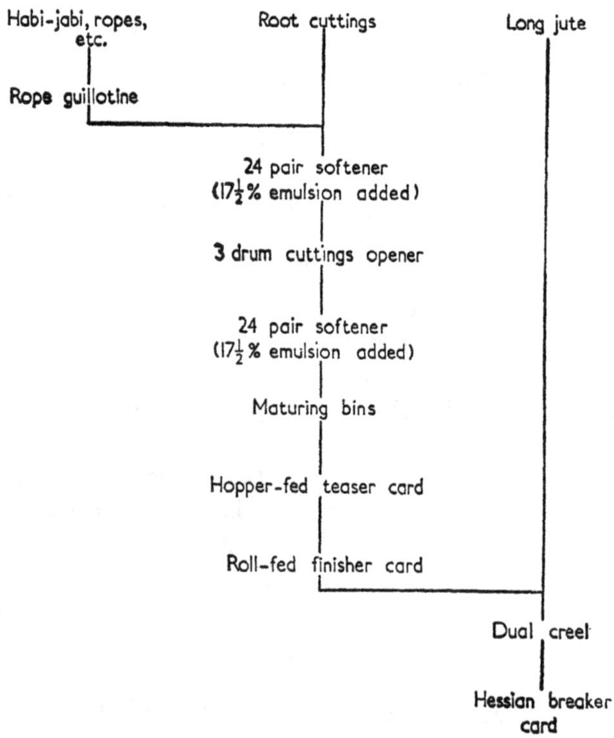

Figure 6.4. Flowsheet for blending cuttings and long jute on the Mackie system

feed sheet at regular intervals across its width and superimposed on the long jute slivers. In this way an intimate blend can be achieved.

The use of these special ranges of machinery permits a more regular product to be made from low grade material. This has enabled an extension of the scope of blending to reduce the batch cost and, as such, represents an important contribution to processing economics.

Figure 6.4 shows one possible arrangement for mixing low-grade material with long jute.

CARDING CALCULATIONS

These are of a simpler nature than those required for the spreader, being confined chiefly to draft, count, and speed. The following three are typical of those met with in practice.

(1) Eight ends of 300 ktex spreader sliver are fed to a breaker with a draft constant of 500 which has a draft change pinion of 28 fitted. If the feed speed is 3 m/min find the card production in an 8 hr day when it runs at 80 per cent efficiency, and the deliver sliver count when there is a moisture and waste loss of 6 per cent of the input weight. The sliver is delivered by a roll-former which has a 4 per cent lead over the delivery rollers of the card.

Draft on the card $= \frac{500}{28}$

$= 17{\cdot}85$

Delivery speed of card $= 3 \times 17{\cdot}85$

$= 53{\cdot}5$ m/min

Therefore, roll former delivery speed $= 53{\cdot}5 \times 1{\cdot}04$

$= 55{\cdot}6$ m/min

Daily delivery $= 55{\cdot}6 \times 60 \times 8 \times 0{\cdot}8$

$= 21400$ m

Sliver count at roll former $= \frac{300 \times 8 \times 0{\cdot}94}{17{\cdot}85 \times 1{\cdot}04}$

$= 121{\cdot}5$ ktex

Daily production $= \frac{121{\cdot}5 \times 21400}{1000}$

$= 2600$ kg

(2) It is necessary to produce breaker card sliver at a count of 18 lb/100 yd. What dollop weight must be used to meet the following conditions?

Draft constant	440
Draft pinion	30
Feed sheet roller	7 in. diameter
Feed sheet roller	24·5 revolutions per clock revolution

Moisture and waste loss 3·5 per cent

In 1 revolution of the clock the feed sheet travels

$$\frac{24{\cdot}5 \times 7 \times 3{\cdot}1416 \text{ yd}}{36} = 14{\cdot}95 \text{ yd}$$

i.e. the clock length is 14·95 yd

$$\frac{\text{Dollop} \times 100}{\text{Clock length} \times \text{draft}} = \text{Delivered sliver (lb/100 yd)}$$

$$\text{Draft} = \frac{440}{30} = 14{\cdot}7$$

Hence, ignoring losses for the moment,

$$\text{Dollop weight} = \frac{18 \times 15 \times 14{\cdot}7}{100} = 39{\cdot}7 \text{ lb}$$

To allow for losses this must be increased by 1·035,

Correct dollop weight = 41 lb

(3) The cylinder of a finisher card rotates at 180 r.p.m. Find its linear speed when its radius is 24 in. The feed roller travels at $\frac{1}{200}$ of the cylinder speed, the doffer at $\frac{1}{40}$ of the cylinder speed and at half the speed of the delivery rollers. Find the card draft.

$$\text{Cylinder surface speed} = \frac{180 \times 2 \times 3{\cdot}1416 \times 24}{12} = 2290 \text{ ft/min}$$

$$\text{Therefore, feed roller surface speed} = \frac{2290}{200} = 11{\cdot}45 \text{ ft/min}$$

$$\text{Similarly, doffer surface speed} = 57{\cdot}3 \text{ ft/min}$$

$$\text{Therefore, delivery speed} = 57{\cdot}3 \times 2 = 114{\cdot}6 \text{ ft/min}$$

$$\text{Machine draft} = \frac{114{\cdot}6}{11{\cdot}45} = 10$$

CHAPTER SEVEN

Drawing

THE functions of the drawing stages are (1) Drafting the finisher card sliver to a count suitable for feeding the spinning frames; (2) Reduction of weight irregularities by doubling; (3) Straightening the fibres and laying them along the sliver axis so that when they come to be spun on the spinning frame they will be evenly drafted and twisted to form an acceptable yarn.

DRAFTING

To examine the behaviour of the fibres during drafting, the simplest case will be considered first, where there are two sets of rollers involved—a feed pair and a drawing pair. The jute sliver enters the machine through the nip of the feed or retaining rollers and then passes forward to the drawing rollers. Because of the greater linear speed of the latter, the material becomes drafted, the exact amount of drafting being determined by the relative surface speeds of the two sets of rollers. The distance between the two sets of rollers, the reach, is longer than the fibre being drafted; if it were not so then a number of fibres would be gripped by both sets of rollers at the same time and be broken. As a result of this comparatively long distance there is always a large number of fibres which are gripped neither by the retaining rollers nor the drawing rollers. These are called 'floating' fibres. For ideal drafting each fibre should move with the same speed as the back rollers until their leading ends enter the nip of the drawing rollers. Under these conditions the fibre tips in the drafted sliver would be Dx inches apart if the draft on the frame was D and the tips were x inches apart in the entrant sliver. In practice, however, a floating fibre is held *in situ* by entanglement with its neighbours and inter-fibre frictional forces. When a long fibre has its tip gripped by the drawing rollers it immediately accelerates to the speed of these rollers and, because of this fibre entanglement and inter-fibre friction, some of the short fibres lying alongside will be dragged forward and prematurely drafted. This process is cumulative so that a clump of short fibres is drafted too soon, producing a thick place in the sliver. Moreover, this action causes a deficiency of floating fibres in the drafting zone with the result that a thin place follows on after a thick place. Such a cycle is repeated as more floating fibres are fed through the nip of the retaining rollers.

This is a simplified picture of drafting with only two sets of rollers but it should be sufficient to show that if no attempt is made to control the movement of the floating fibres the resulting sliver will be highly irregular. In jute drawing frames short fibre control is obtained by means of moving sheets of pins which carry the sliver up to the nip of the drafting rollers. The pins provide sufficient restraint to stop most of the short fibres being drafted prematurely but at the same time do not interfere with the normal processes of drafting near the nip of the drawing rollers. These types of drawing frames are known as gill-drawings and the pins as gill-pins.

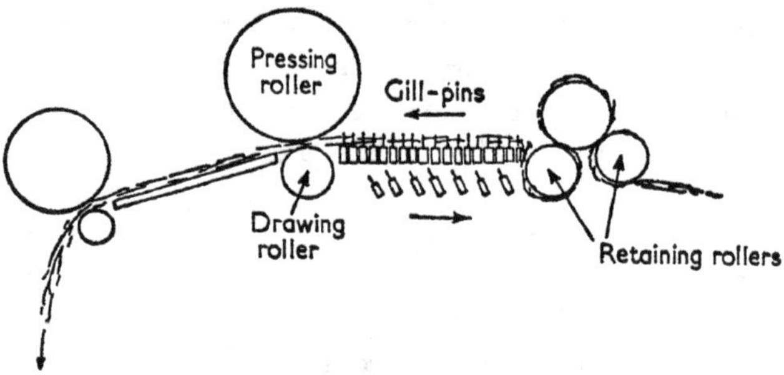

Figure 7.1. General outline of a drawing frame

Figure 7.1 shows the general lay-out of the drafting mechanism of a jute drawing frame. The slivers enter the machine between the retaining rollers and a self-weighted jockey-roller and then meet the gill-pins. The gill-pins are carried on a series of faller-bars which move in the direction indicated by the arrows. As the sliver leaves the nip of the back rollers a faller-bar with its sharp pins strikes upwards into it and the fibres are impaled on the gill-pins. The faller-bars move forward as a sheet and carry the sliver to the front of the machine. When the faller-bars are close to the drawing rollers they drop out of the sliver and travel back underneath the sliver in preparation for another strike upwards. The relative surface speed of the drawing and retaining rollers determines the draft in the normal manner. The linear speed of the gill-pins is a few per cent higher than that of the retaining rollers so that when the sliver is held between the retaining rollers and the pins that have just struck into it, it is under slight tension and the next row of pins can penetrate the sliver more easily. It is essential that the sliver rides within the pins otherwise control over the short fibres will be lost; if the sliver lies on top of the pins it is equivalent to drafting with only two sets of rollers and no draft control mechanism there at all. The sliver will ride on top of the pins when the lead between the retaining roller and the faller-bars is too low, or when the weight of the sliver is too great for the machine, or when the pins are blunt and hooked at their tips instead of being keen and sharp. The first essential of good draft control is good pinning.

Ideally, the pins should accompany the sliver right up to the nip of the drawing rollers so that even the very shortest fibres are controlled until the last minute but because of the dimensions of the bars and the

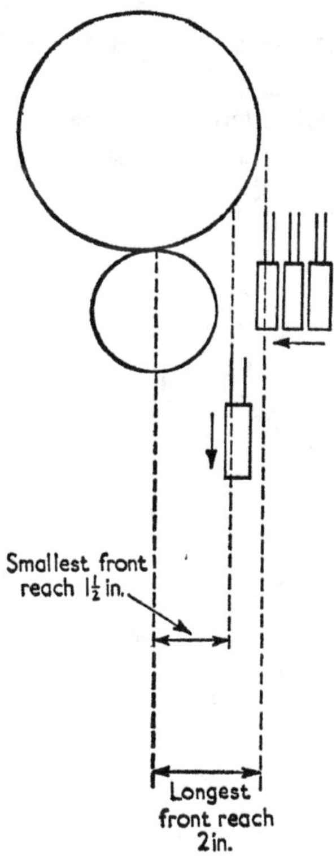

Figure 7.2. Drawing frame front reach

rollers this is not possible and there is a gap with no fibre control at all just at the most critical zone in the whole drafting area. Figure 7.2 shows how the distance between the point at which the faller-bars must drop out and the drawing roller is determined by the drawing roller diameter and the thickness of the faller-bar. While the machine is running, this distance, the front reach, varies between two extremes depending on the pitch of the faller-bars. For instance, in Figure 7.2, if the diameter of the drawing roller is $2\frac{1}{2}$ in. then the pins on the faller-bar may not be able to approach nearer than, say $1\frac{1}{2}$ in. to the nip of the drawing rollers. This means that there is always at least $1\frac{1}{2}$ in. of uncontrolled sliver. In the Figure, one bar has just dropped out of the sliver and therefore if the pitch of the pins is, say, $\frac{1}{2}$ in. then this uncontrolled length is now increased to 2 in. $(1\frac{1}{2}+\frac{1}{2})$. The result of this variable distance front reach is to allow uncontrolled drafting and, with a material like jute where there are many extremely short fibres in the sliver, the formation of what may be termed faller-bar drafting waves, or more simply faller-bar slubs. ('Slub' is a general term used to denote a thick clump of fibres in a sliver or a yarn.)

Figure 7.3 shows the fibre length distribution of a jute finisher card sliver placed alongside the drafting zone of a first drawing frame and

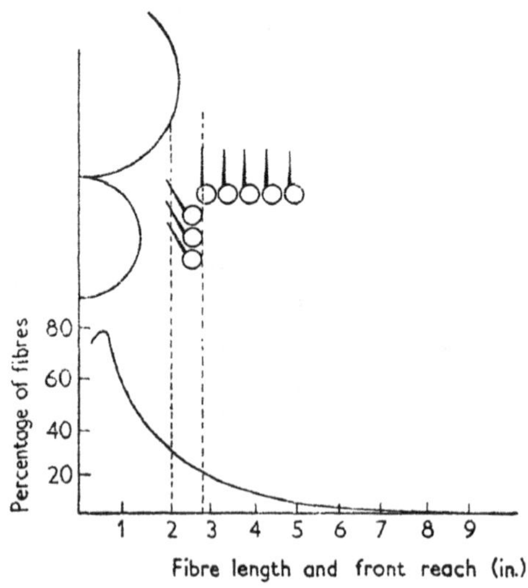

Figure 7.3. Number of finisher card slivers shorter than the front reach on a push-bar first drawing frame

the high proportion of fibres which are shorter than the front reach will be seen. Naturally, the more short fibre present in a sliver the more fibre movement will occur in the front reach and the more pronounced the faller-bar slubs will be.

The wavelength of the faller-bar slubs can be found from the product of the draft and the pitch of the faller-bars, e.g. with a draft of 4 and fallers $\frac{3}{8}$ in. apart, the faller-bar wavelength will be $1\frac{1}{2}$ in. Figure 7.4 shows sliver irregularity charts obtained from a testing machine which 'weighs' a continuous length of sliver electronically; in this test each alternate faller-bar was deliberately removed to illustrate the effect of the front reach. The weight profile of the sliver shows

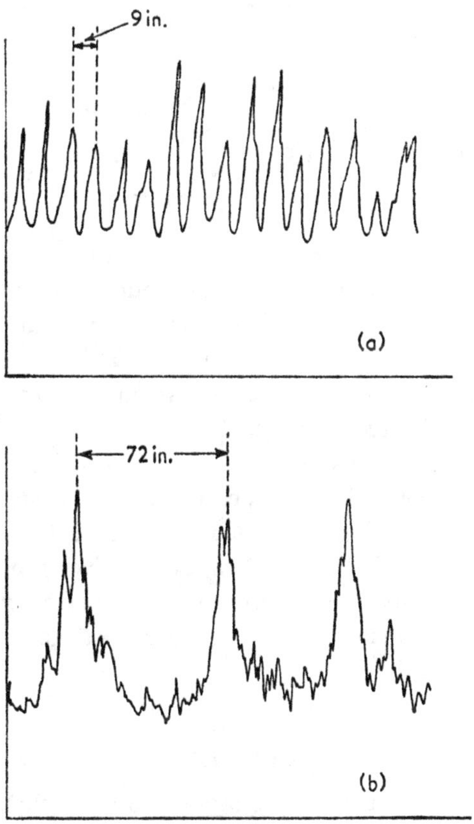

Figure 7.4. Faller-bar slubs in sliver and yarn:
(a) *Weight profile of sliver when alternate faller-bars are removed;*
(b) *Weight profile of yarn spun from above sliver at a draft of 8*

the accentuated faller-bar slub and the yarn trace shows that the slubs are carried forward to the finished yarn.

Faller-bar slubs cannot be eliminated entirely but they can be kept small by working with the correct density of pinning. As the sliver progresses through the drawing stages and its count becomes smaller then the gill-pins require to become finer and more closely-set, as is shown by Table 7.1

TABLE 7.1. DRAWING FRAME PINNINGS

	First push-bar	*Intermediate spiral*	*Finisher spiral*
Faller pitch (in.)	$\frac{1}{2}$	$\frac{1}{2}$	$\frac{3}{8}$
Pins/inch on bar	2	5	8
Rows of pins on bar	1	2	1
Width of gill (in.)	6	5	$3\frac{3}{4}$
Pin w.g.	13	15	15
Length of pins (in.)	$1\frac{1}{8}$	1	$\frac{7}{8}$

Drafting, therefore, is closely connected with sliver regularity and because of the variability of fibre length found in jute slivers and the imperfections of the draft control mechanisms each drafting operation increases the amount of irregularity present. This it does in two ways. If the entrant sliver contains a weight irregularity with a wave-length of 4 in. and it is drafted 4 times, then the delivered sliver will have an irregularity with a wave-length of $4 \times 4 = 16$ in. In addition, the action of drafting will have imposed further irregularities on the material and while the basic wave-length may be 16 in. there will be minor irregularities added to it. Thus, as the sliver progresses through the drawing stages its irregularity pattern becomes more and more complex and it is the function of doubling to try to reduce this complexity as far as possible.

DOUBLING

In jute slivers the count varies from place to place along the length of each strand and there are also differences in the general count level from one sliver to another. These differences in count fall into a definite pattern which can be defined statistically by the Normal

Distribution. If one took a length of sliver and cut it up into sections of, say, 1 ft and weighed them, one would find that the distribution of count followed a bell-shaped pattern (it would require a large number of tests to arrive at a smooth curve but even about 100 results show the general pattern). Figure 7.5 (a) shows the type of result obtained from such a test.

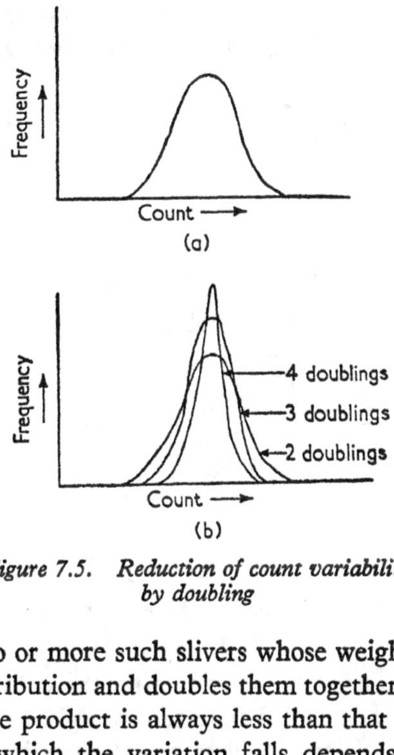

Figure 7.5. Reduction of count variability by doubling

If one takes two or more such slivers whose weight varies according to the normal distribution and doubles them together then the variation in the count of the product is always less than that of the individuals. The amount by which the variation falls depends upon how many slivers are doubled together, in fact the variation falls according to the square root of the number of doublings. Figure 7.5 (b) shows the distribution of count after 2, 3, or 4 slivers have been doubled together (the narrower and taller the bell-shape, the better and more uniform is the sliver). Doubling, it may be seen, is highly advantageous, but it should not be forgotten that to reduce the count of the material drafting must predominate over doubling and that drafting increases the variability of weight. Therefore, as the material passes over the drawing stages there are the two conflicting influences at work; doubling leading to a greater uniformity of weight and drafting leading to greater irregularity of weight.

Doubling may be carried out by placing two or more slivers together at the feed end of the machine and entering them on to one set of gill-pins or by uniting the slivers as they emerge from the nip of the drawing rollers in which case there is only one sliver on each set of pins in the drafting zone. The former situation holds for the lighter counts of sliver at the last drawing passage but in the earlier ones only one sliver can be accommodated on the pins and doubling takes place at the front of the machine. There is, however, another reason for doubling at the front of the machine and this is connected with the faller-bar slubs. In Figure 7.6 the plan view of a first drawing frame

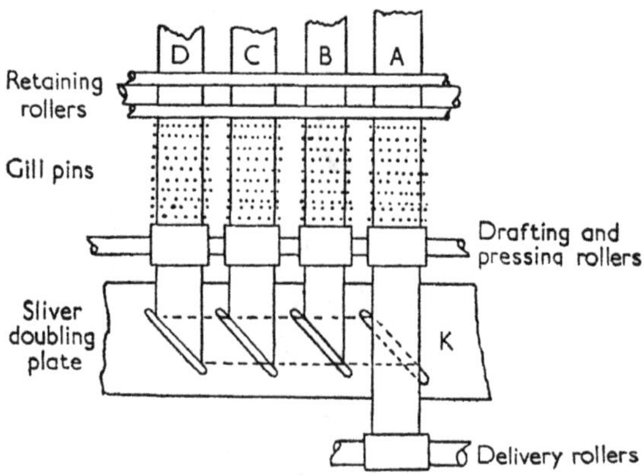

Figure 7.6. 4–1 doubling on the sliver doubling plate

with four doublings has been shown. After the individual slivers have been drafted they are doubled together on a plate between the drawing rollers and the delivery rollers called the sliver doubling plate. This is a cast iron plate roughly 1 in. in section running across the front of the machine. Slots with rounded edges are cut in the plate at an angle of 45 degrees to the line of the frame, through which the slivers can pass so as to change their direction. In the set of four doublings one sliver comes straight out of the drawing nip towards the delivery rollers but the other three are turned through 45 degrees and pass along the back of the plate to another 45 degrees slot. When they pass through this second slot they are laid down on top of one another and are now travelling toward the delivery rollers. The four doubled slivers now pass through the delivery nip where they are consolidated into one sliver and leave the machine.

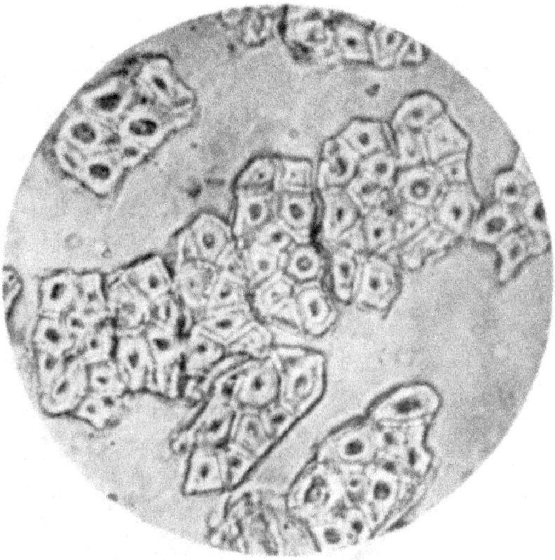

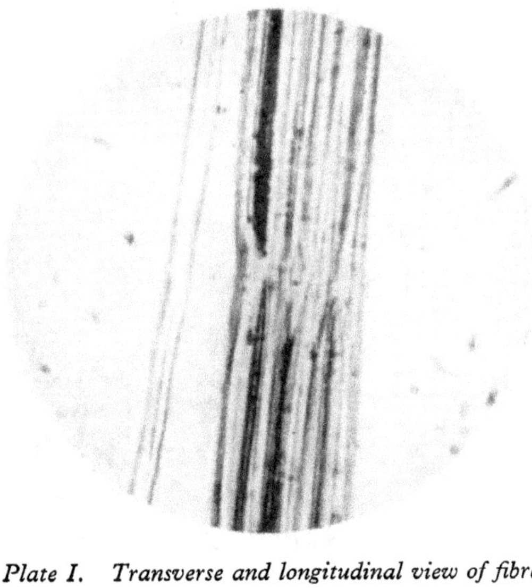

Plate I. Transverse and longitudinal view of fibre

Plate II. Root cuttings sliver dispersal unit

Plate III. Dual creel for blending long jute and cuttings sliver

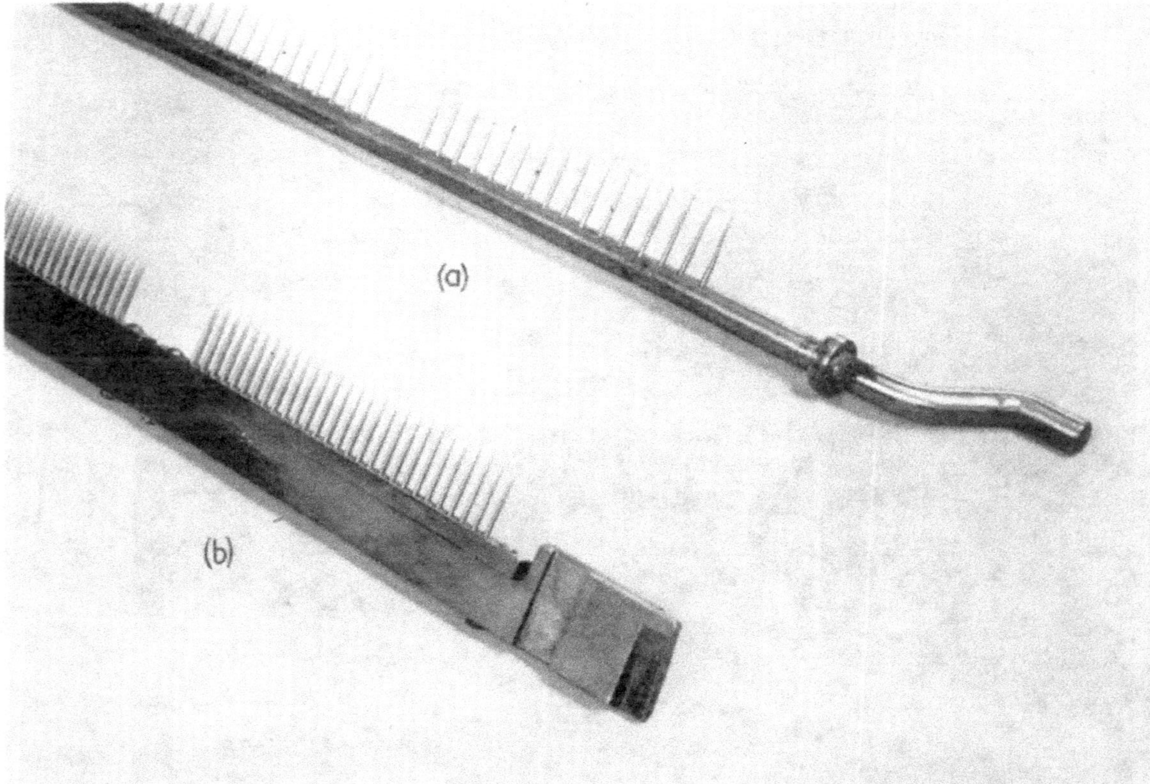

Plate IV. Drawing frame fallers: (a) *push-bar,* (b) *spiral*

Plate V. Methods of leading yarn on to the bobbin

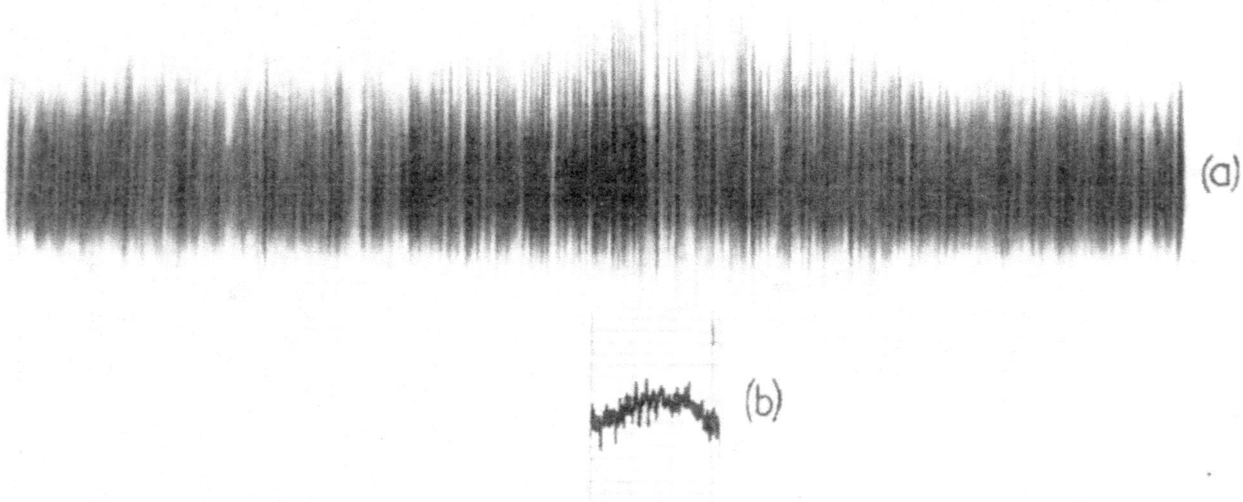

Plate VI. Spindle vibration and its associated tension peaks:
(a) *Spindle vibration;* (b) *Tension trace*

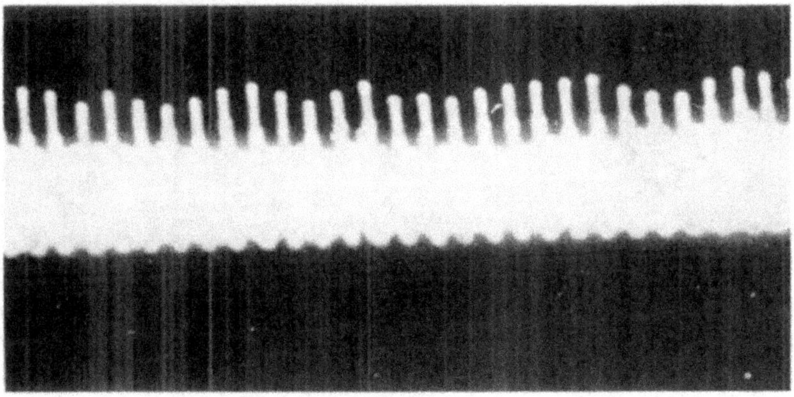

Plate VII. Transmitted tension pulses due to tape joint

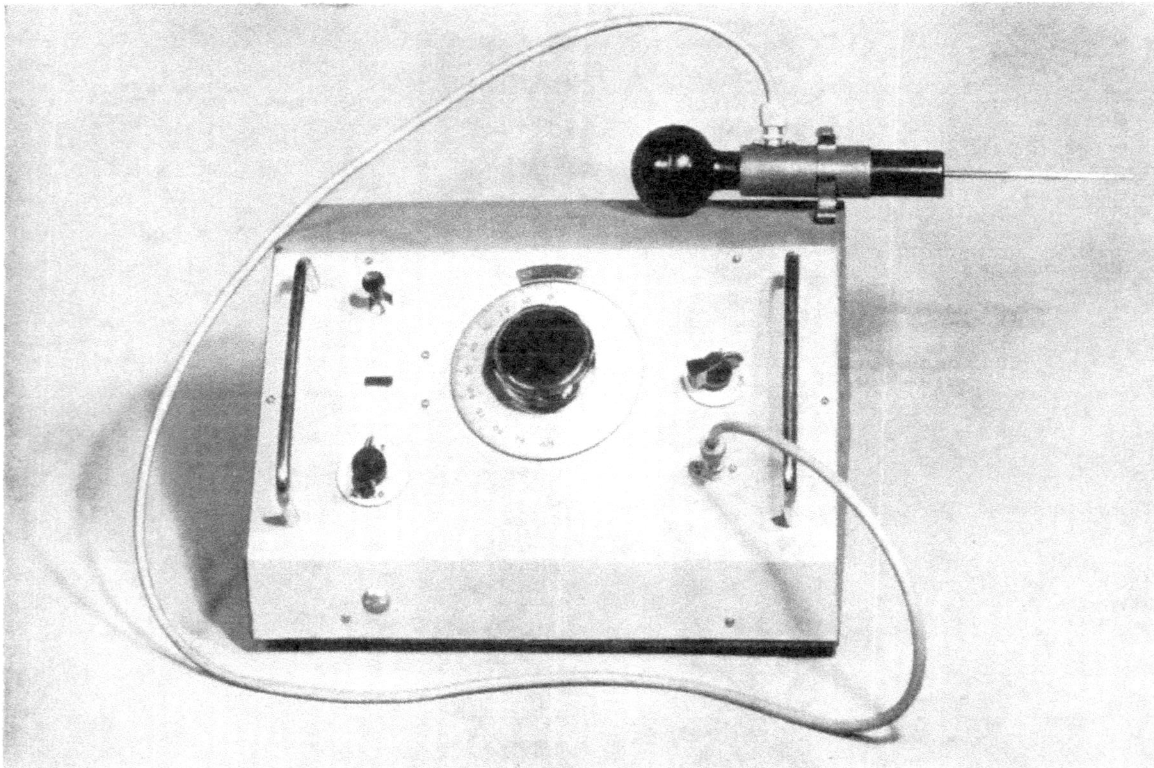

Plate VIII. B.J.T.R.A. Probe Moisture Meter

In order to examine the working of the doubling plate it will be assumed that the sliver entering the drawing frame has been perfectly uniform. As the four slivers traverse the gill-sheet they are held in the same way and when the faller-bars drop out of the sliver a faller-bar slub appears in each at exactly the same point. There are now four slivers with identical wave-forms issuing from the drafting nip, peak with peak, trough with trough, in perfect phase. Ultimately, from the sliver paths on the doubling plate, these four slivers are going to be placed one on top of the other in a four-layer sandwich. If all the peaks in the slivers coincide then the resulting sliver will be extremely irregular but if peaks can be made to fall alongside troughs then a more uniform product will result. The combination of the sliver doubling plate design, the draft, and the pitch of the faller-bars decides which of these conditions will prevail.

A numerical example will perhaps help to clarify this statement. Consider the doubling plate shown in Figure 7.6, and suppose that the four slivers are issuing from the drawing rollers with a faller-bar slub wave-length of 2 in. (this could arise from a faller pitch of ½ in. and a draft of 4 on the frame). The slivers unite as a point K. Sliver A has a path length of 8 in. and since the wave-length is 2 in. there will be $8/2=4$ complete waves in this length of sliver. Sliver B has a path length of 16 in. in which there will be 8 complete waves, C has a path of 24 in. with 12 waves in it, and D has a path of 32 in. containing 16 waves. Thus at the uniting point, K, the four slivers will come together with the peaks of each wave-length coincident and the resulting sliver will be more irregular than either sliver A, B, C, or D.

If, on the other hand, the wave-length is 2½ in. (resulting from a faller pitch of ½ in. and a draft of 5) sliver A will have $8/2{\cdot}5=3{\cdot}2$ waves, sliver B 6·4 waves, C 9·6 waves, and D 12·8 waves. At the point K the peaks of the waves will be 0·2, 0·4, 0·6, and 0·8 wave-lengths apart and this will have the effect of producing a more regular sliver in which the ill-effects of the faller-bar slubs have been reduced to a minimum.

The faller-bar pitch and the path length on the doubling plate are fixed by the machine designer and the only variable left under the control of the producer is the draft. When the draft is changed the faller-bar slub wave-length is altered and hence the number of wave-lengths on the sliver doubling plate is changed, as in the example above. If the pitch of the slots in the doubling plate is P, the faller bar pitch p, and the draft d, then for any number of doublings the 'worst' draft, i.e. that one leading to peak-on-peak doubling, is given by,

$$d = \frac{P}{np}$$

where n is any whole number.

For two or more doublings, the 'best' draft, i.e. giving peak-on-trough doubling, is given by

$$d = \frac{P}{(n+\frac{1}{2})p}$$

and for three doublings

$$d = \frac{P}{(n+\frac{1}{3})p}$$

The implication of these relationships is important, for once the number of doublings has been chosen, the choice of drafts available is fixed by the design of the doubling plate. Certain drafts will produce more irregular material than others simply because they impose peak-on-peak doubling on the sliver instead of peak-on-trough doubling.

The practical results of working with a 'good' draft on a drawing frame may be illustrated by the following yarn test figures obtained when the second (intermediate) drawing frame was run at the draft which laid peaks and troughs together and another which superimposed the peaks on peaks.

	'Best' draft	*'Worst' draft*
Count (tex)	520	520
Tenacity (g/tex)	13·4	11·7
Minimum tenacity (g/tex)	8·0	6·0
Short term weight irregularity (per cent)	20·0	22·0

There is a difference in the average tenacities but the important practical value in a yarn is the strength of its weakest point and in this test the minimum tenacity (allowing for the normal variations in yarn strength) was some 33 per cent higher when the 'best' draft was used. Though this was a laboratory-scale trial similar results have been found in normal mill conditions and clearly the benefits accruing from the proper choice of draft are well worth seeking.

TYPES OF DRAWING FRAME

Jute drawing frames are divided into two types, depending on the mechanism used to propel the faller-bars.

(1) *Push-bar.* In this class, the fallers have specially cranked ends which run in slides on the machine frame. The fallers are driven by a large carrier wheel at the back of the machine. The earlier models had collars on each faller-bar which bore against each other but in modern frames the bars bear across the full width, the bar behind pushing the bar in front—hence the name.

(2) *Spiral.* In this method of faller-bar propulsion there are two spiral screws on each side, one set directly above the other. The ends of each faller-bar are cut to fit into the grooves on the spiral so that as the screws rotate they drive the faller-bars along. As each faller comes to the end of the top screw it is knocked down on to the bottom one by a cam on the top screw, springs holding it steady as as it falls into the grooves of the bottom screw. The bottom spiral is more coarsely pitched than the top one so that the faller-bars are returned quickly to the back of the machine ready to be lifted by cams on the bottom screw up into the spirals of the top screw. By having a coarse spiral on the bottom fewer bars are needed to complete the gill sheet.

PUSH-BAR MACHINES

Figure 7.7 shows one type of push-bar drawing frame, and the cranked end of one of the faller-bars is illustrated in Plate IV (a). Each bar is cranked only at one end and the carrier wheel has half as many teeth as there are faller-bars, alternate bars being driven from opposite sides of the machine.

In addition to being driven round the machine the bars must present

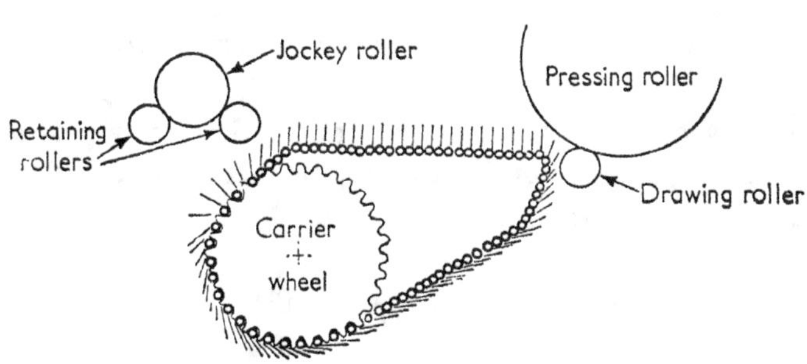

Figure 7.7. Push-bar frame

their pins to the sliver in as advantageous a manner as possible. This requires that the pins shall enter the sliver cleanly and show little tendency to lift the material on their points rather than pierce it, and that they shall leave the sliver without drawing down loose fibres. To achieve these functions there are tracks for guiding the bars in their course around the machine. The first of these is the guide track which keeps the faller bars in the correct position as they travel round. The other tracks, the pin control tracks, ensure that the pins enter and leave the sliver in the desired manner. When the pins are about to enter the sliver the pin control tracks, by virtue of their position relative to the guide tracks, act on the cranks in such a manner that the pins are swung into a vertical position, ready for a clean strike into the sliver. At the draft end of the machine the pin control tracks force the bars to change their orientation so that the pins fall freely from the sliver by swinging forward.

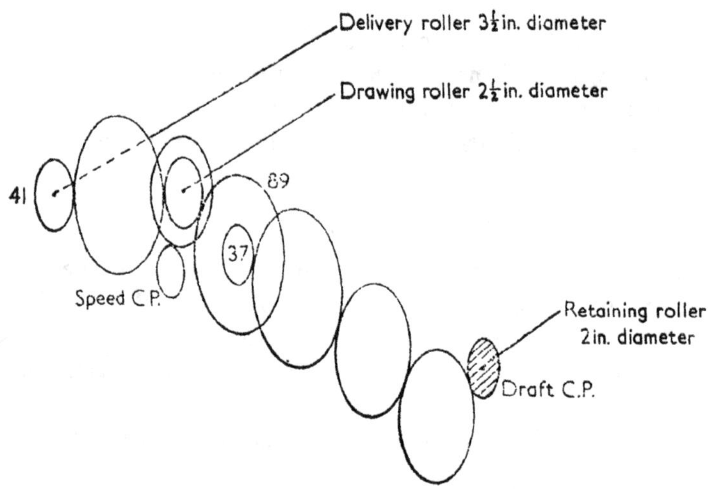

Figure 7.8. Draft gearing in a push-bar drawing frame

Figure 7.8 shows a diagram of the draft and delivery gearing of a push-bar drawing frame. For illustration, the method of calculating the draft constant is shown:

$$\frac{1}{2 \text{ in.}} \times \frac{\text{c.p.}}{37} \times \frac{89}{41} \times \frac{3\frac{1}{2} \text{ in.}}{1} = 0{\cdot}1 \times \text{c.p.}$$

i.e. $\quad$ draft $= 0{\cdot}1 \times$ draft change pinion

Similarly, the lead of the delivery rollers is calculated by working from the slower roller forward to the faster roller and then expressing the lead of the latter as a percentage

$$\frac{1}{2\frac{1}{2}\text{ in.}} \times \frac{30}{41} \times \frac{3\frac{1}{2}\text{ in.}}{1} = 1{\cdot}0244$$

i.e. the faller lead is 2·44 per cent

SPIRAL DRAWING FRAMES

Plate IV (b) shows one end of a faller-bar from a spiral drawing frame and the screws which carry it are illustrated in Figure 7.9. Modern spiral frames are all double-thread or triple-thread, i.e. there are two

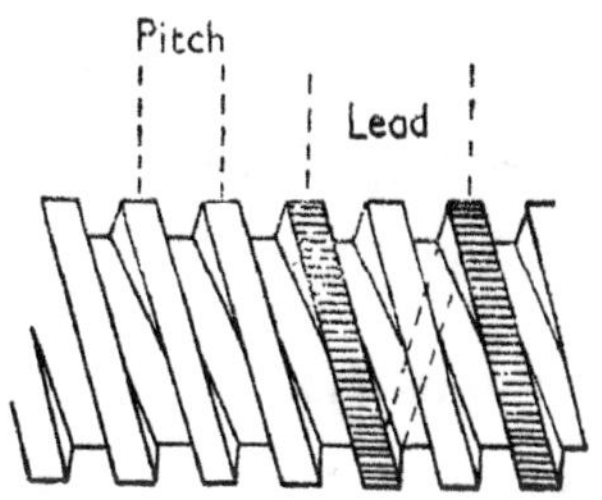

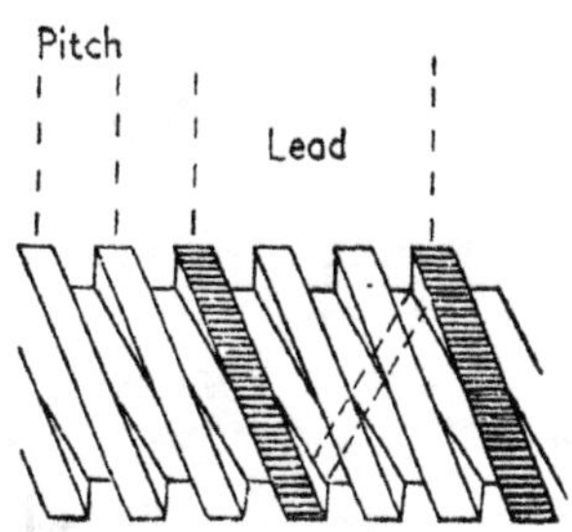

Figure 7.9. Double- and triple-thread screws for spiral drawings. (a) *Double-thread, one revolution of screws moves fallers a distance equal to the lead, i.e., 2 × pitch;* (b) *Triple-thread, one revolution of screw moves faller a distance equal to the lead, i.e., 3 × pitch*

or three complete spirals cut in each screw. The length of one complete spiral is the lead and the distance between adjacent spirals, the pitch.

$$\text{Faller-bar speed} = \text{r.p.m. of screw} \times \text{lead}$$
$$\text{Lead} = \text{number of screws} \times \text{pitch}$$

Early models of spiral frames had single-thread screws and the introduction of the double- and triple-thread has allowed faller-bar speeds to be greatly increased, as the limiting factor in a spiral frame is the rate at which the fallers can be dropped out of and lifted into the sliver. If the speed is increased above 200 drops per minute with a single thread spiral the bars begin to jump, pin badly, and the wear and tear is high, but with double screws faller drops of about 400 per minute are possible and with triple screws about 650.

The general lay-out of a spiral drawing frame was shown in Figure 7.1.

On this type of frame the pins are usually carried on brass gill-stocks which are riveted on to the bar. These make for easy pin renewal.

COMPARISON OF PUSH-BAR AND SPIRAL MACHINES

The following Table gives a condensed comparison between the two types of drawing frames.

Push-bar	*Spiral*
Faller drops up to 850/min	Double screw up to 400 drops/min Triple screw up to 650 drops/min
Faller-bar lead over retaining rollers 4–10 per cent	Faller-bar lead $1\frac{1}{2}$–$4\frac{1}{2}$ per cent
Quiet running	Noisy
Tends to clog with dirt	Self cleaning because of the jerk at each drop
Pinning good with modern types	Pinning excellent
Laps occasionally, especially with light slivers	Seldom laps

Since the object of any industrial process is to achieve a high production rate at an acceptable quality level as economically as possible it is desirable to be able to run machinery at high speeds. The highest speeds at which gill drawing frames can be operated is given by the equation

$$f = \frac{v}{pd}$$

where f is the number of faller drops per minute, v is the delivery speed per minute, and d is the machine draft.

In practice there is an upper limit to f, the faller drops per minute, imposed by two factors. The first of these is the ability of the fallers, as machine components, to withstand the forces involved in their propulsion without excessive amounts of wear and tear, the second factor is the ability of the gill-pins to strike into the sliver and control fibre movement during drafting. With regard to the mechanical aspects the fallers on spiral drawing frames, with their sudden drop-out at the drawing rollers and their rise at the feed rollers are subjected to greater strains than those of push-bar machines and for this reason cannot achieve such high speeds as the push-bar types. The pitch of the faller-bars is closely related to the maximum speed of the gill-sheet, for the smaller the pitch p, the finer must be the bars, pins, screws, etc., and the more expensive the mechanism becomes. There is obviously a lower

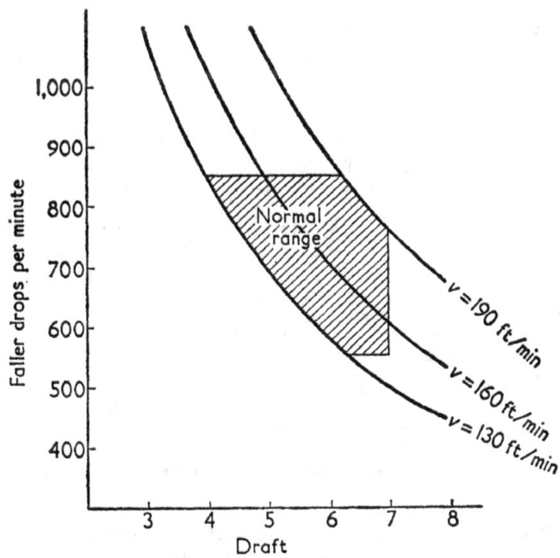

Figure 7.10. Effect of drawing draft and delivery speed on faller drops per minute. Sacking weft, 1st push-bar frame $f = v/dp$

limit beyond which the materials used and the manner of construction become so refined that the cost becomes prohibitive. On jute frames the faller bar pitch is between $\frac{3}{4}$ and $\frac{3}{8}$ in. and, as far as the user is concerned, this can be regarded as being fixed by the machine designer. Control over the number of faller drops per minute, therefore, devolves on making adjustments to the speed of the whole machine by a series of speed pinions on the main drive and selecting a suitable draft. Figure 7.10 shows how draft and delivery speed combine to give a series of different faller drops per minute on a sacking weft push-bar first drawing frame. In this case if it is desired to work at the upper limit of faller drops for this type of machine (850 per minute) and the faller pitch is 0·5 in., then,

$$v = 425d \text{ in./min}$$

and if the draft is changed at any time the delivery speed should be altered also to ensure that this relationship holds and the machine is run at its maximum speed compatible with freedom from mechanical trouble and correct pinning of the sliver.

DRAWING SYSTEMS

The common arrangement for hessian qualities is to have three drawing passages over a first push-bar, a second (or intermediate) double-thread spiral, and a finisher triple-thread spiral drawing. A double-

TABLE 7.2. EXAMPLES OF DRAWING SYSTEMS. ALL SLIVER COUNTS IN LB/100 YD

	(1)	(2)	(3)	(4)	(5)
Finisher card sliver count	14·0	15·0	16·0	9·0	18·0
First drawing draft	4·7	4·0	3·5	4·6	5·0
First drawing sliver count	12·0	15·0	9·1	5·9	7·2
Intermediate drawing draft	7·0	6·5	6·0	—	—
Intermediate drawing sliver count	5·2	4·6	4·5	—	—
Finisher drawing draft	10·0	9·0	9·0	10·0	7·5
Finisher drawing sliver count	1·04	1·02	1·0	1·18	1·90

Key:

(1) Hessian system, 4-3-2 Doublings, three drawing passages.
(2) Hessian system, 4-2-2 Doublings, three drawing passages.
(3) Hessian system, 2-3-2 Doublings, three drawing passages.
(4) Hessian system, 3-2 Doublings, two drawing passage.
(5) Sacking Weft system, 2-2 Doublings, two drawing passages.

thread spiral frame may be used as a first drawing where better quality work is desired but its speed and production are not so high as those of a push-bar and so more machines are required to handle the same quantity of fibre.

Certain hessian and sacking warp systems have only two drawings, working in conjunction with a drawing head on the finisher card to reduce the sliver count. Sacking weft systems have only two drawing passages in order to keep the manufacturing costs as low as possible.

Table 7.2 shows examples of several drawing systems with different numbers of doublings at the first and intermediate drawing stages.

From the data in Table 7.2 it is possible to analyse these systems from the point of view of the number of doublings and the stages at which most mixing occurs. The number of doublings in a system is found by *multiplying* together the doublings at each stage, e.g. 4 doublings at the first, 3 at the intermediate, and 2 at the finisher drawing frames gives a total of $4 \times 3 \times 2 = 24$ doublings. If the net draft at each stage is calculated, i.e. (machine draft)/(doublings), then the closer this is to 1 the more mixing and evening out of irregularities is occurring.

TABLE 7.3

	Drawing systems				
	(1)	(2)	(3)	(4)	(5)
Total number of doublings	24	16	12	6	4
Net draft at first drawings	1·17	1·00	1·75	1·50	2·50
Net draft at intermediate drawings	2·31	3·33	2·00	—	—
Net draft at finisher drawings	5·0	4·6	4·5	7·5	3·8

Notice that in the hessian systems the first drawing net draft is between 1 and 2, indicating that this stage is used primarily as a doubling stage. Most of the attenuation occurs at the final drawing stage. The number of drawing stages adopted depends upon the count and quality range to be spun, the nature of the raw material, the efficiency of the draft control mechanisms and, of course, process cost and labour requirements. If heavy sacking yarns are to be made then a two-drawing system will be chosen but if 4–6 lb/sp yarn of top quality is required then 4 drawing passages will be needed since high quality demands many doublings and short drafts. The shorter the drafts and the more doublings there are, the costlier is the process and, as so often happens in industry, a compromise must be reached between the demands of quality, production, and cost.

CRIMPED SLIVER

As the count of the sliver is reduced in its passage through the drawing stages it becomes more and more fragile until, by the time it emerges at the finisher drawing delivery, it is in so tenuous a form that it is impossible to handle at all and, indeed, is so weak that it would not carry up the back of the spinning frames. To overcome this, the sliver is crimped, or waved, to give a certain amount of cohesion to the strand. In some drawing systems the sliver at the first and intermediate drawing frames is crimped but all systems use crimped finisher drawing sliver. Figure 7.11 shows a crimping box attached

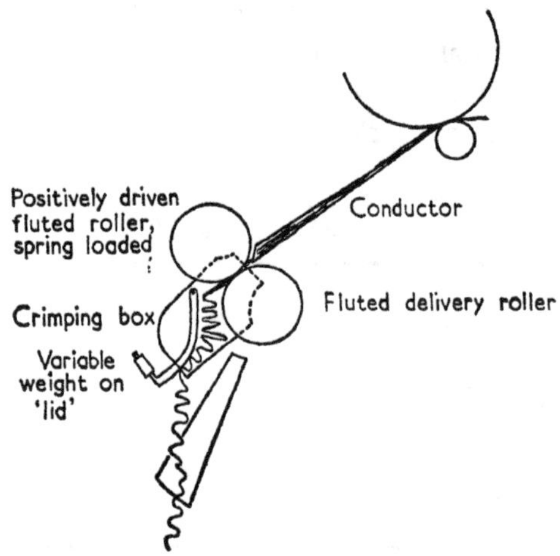

Figure 7.11. Crimping box attachment on a finisher drawing frame

to the delivery of a finisher drawing frame. The sliver leaves the nip of the drafting rollers and passes down the sliver plate into the nip of a pair of fluted delivery rollers, the upper one of the pair being spring-loaded and positively driven through a wide-pitch gear from the lower one. The sliver is driven into the box where it meets a metal finger or lid hanging down into the box. The finger impedes the motion of the sliver and the box quickly fills, when more sliver enters at the back the lid of the box is forced up by the mass of sliver inside the box and the sliver at the front of the box can come out; this, of course, is a continuous process, although the delivery of the crimped sliver is not steady and the sliver spurts out at an irregular rate from second to second. During its sojourn in the box the fibres in the sliver become 'concertinad' and take on a permanent crimp or wave. The length of time any particular piece of sliver remains in the crimping box can be regulated by means of small weights which can be added to the finger, a heavy weight requiring a greater mass of sliver in the box to lift it up and, hence, developing greater crimp in the fibres.

SLIVER PACKING

First and intermediate drawing sliver may be packed in rolls on roll-formers similar to those found on cards or, alternatively, in cans. It may be mentioned that if the sliver at either of these frames is crimped then it must be put into cans—the action of roll-forming would remove most of the crimp. The sliver from the finisher drawing frame is always fed into cans. Common can dimensions are

First drawing	18 in. dia. × 40 in. tall.
Second drawing	14 in. dia. × 40 in. tall.
Finisher drawing	12 in. dia. × 40 in. tall.

In order that the sliver may be packed neatly in the cans and as great a packing density as possible achieved the cans rest on can-turning plates at the front of the machine. These are simply carrier plates which revolve through almost 360 degrees in one direction and then reverse, the cyclic motion coiling the sliver neatly in the can. In addition to these can-turning plates there are a series of can-tramping arms, one for each delivery on the frame. These carry expanded metal 'feet' at their bottom ends, the feet projecting into the cans. As the trampers move up and down they pack the sliver down into the can and allow greater quantities to be inserted.

Automatic stop motions are an essential part of any machine which is meant to have a high output and the minimum of supervision. Jute drawing frames are fitted with a variety of stop motions which will cut off the power supply to the motor if a feed sliver breaks or a lap builds up at the feed or delivery. These devices not only prevent bad sliver being made when, for instance, a feed sliver breaks, but prevent accidental damage to the machine. Another device incorporated to avoid damage to the gill-sheet is the pitch-pin. This is a pin which passes through two flanges on the back-shaft of the machine. The pin, in effect, acts as a coupling between the flanges, transmitting motion from one to the other. If a sudden load is thrown on the faller-bars, perhaps by sliver lapping or choking somewhere, then the pin fractures and the drive to the faller-bars is stopped and damage avoided. It is obvious that the correct type of pin must be used and if a makeshift one is put in which is too strong then the whole object of the safety mechanism is defeated.

CALCULATIONS

The calculations required at the drawing passages are confined chiefly to those concerning sliver count and machine performance. A full set of machine performance calculations will be shown for an intermediate drawing frame of the double-thread spiral variety, Figure 7.12 showing the relevant gearing.

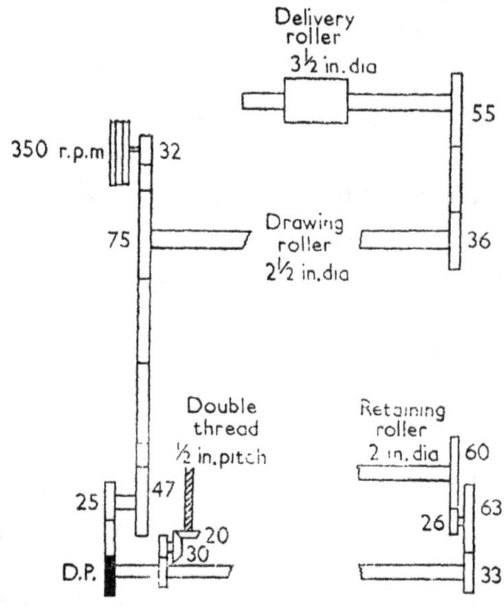

Figure 7.12. Spiral 2nd drawing frame gearing

Drawing roller surface speed:

$$350 \times \frac{32}{75} \times 2{\cdot}25 \times 3{\cdot}14 = 1{,}055 \text{ in./min}$$

Delivery roller surface speed:

$$350 \times \frac{32}{75} \times \frac{36}{55} \times 3{\cdot}5 \times 3{\cdot}14 = 1{,}074 \text{ in./min}$$

Retaining roller surface speed:

$$350 \times \frac{32}{47} \times \frac{25}{\text{c.p.}} \times \frac{33}{63} \times \frac{26}{60} \times 2{\cdot}0 \times 3{\cdot}14 = \frac{8{,}496}{\text{c.p.}}$$

Faller-bar surface speed:

$$350 \times \frac{32}{47} \times \frac{25}{\text{c.p.}} \times \frac{30}{20} \times 0{\cdot}5 \times 2 = \frac{8{,}960}{\text{c.p.}}$$

Faller drops:

$$350 \times \frac{32}{47} \times \frac{25}{\text{c.p.}} \times \frac{30}{20} \times 2 = \frac{17{,}920}{\text{c.p.}}$$

Draft constant:

$$\frac{1}{2 \text{ in.}} \times \frac{60}{26} \times \frac{63}{33} \times \frac{\text{c.p.}}{25} \times \frac{47}{75} \times 2{\cdot}5 \text{ in.} = \text{c.p.} \times 0{\cdot}138$$

i.e. draft = draft change pinion × 0·138

Lead of delivery over drawing rollers:

$$\frac{1}{3{\cdot}5 \text{ in.}} \times \frac{55}{36} \times 2{\cdot}5 \text{ in.} = 1{\cdot}018, \text{ i.e. } 1{\cdot}8 \text{ per cent}$$

Lead of fallers over retaining rollers:

$$\frac{1}{2 \text{ in.} \times \pi} \times \frac{60}{26} \times \frac{63}{33} \times \frac{30}{20} \times 2 \times 0{\cdot}5 \text{ in.} = 1{\cdot}052, \text{ i.e. } 5{\cdot}2 \text{ per cent}$$

DEVELOPMENTS IN DRAWING FRAMES

Much attention has been given in recent years to the possibility of designing a machine which could take account of the irregularities in the sliver as it enters the drawing frame and, by acting on these irregularities by means of a variable draft, produce a regular sliver at the delivery end. That is to say, thick pieces of sliver would be drafted more than thin pieces and the net result would be a greatly improved sliver as far as count regularity is concerned. The first commercially available machine for this was the Raper Autoleveller for worsted slivers and since then many manufacturers have marketed machines for the same purpose. None of these, however, are suitable for jute slivers because of the large variations in weight which are present. In all these machines the drafting mechanism is virtually unchanged, except that an independent variable-speed drive is provided for the drawing rollers; where they differ is in the method adopted for detecting the variations in the feed sliver weight and converting these variations into signals which will be used to control the speed of the drawing rollers. Figure 7.13 illustrates the general principle in such a frame, developed at the B.J.T.R.A.

The sliver passes between one of the normal retaining rollers and another pivoted, counter-balanced roller and, as the bulk of the sliver between the rollers varies, the pivoted detecting roller moves up and down. In this manner the detecting roller follows the weight profile

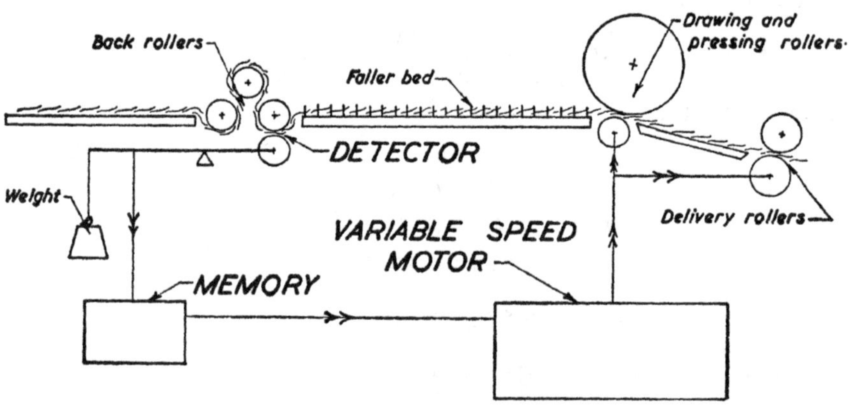

Figure 7.13. Variable draft drawing frame

of the sliver. The variable draft will ultimately operate on the signals put out by the pivoted rollers—increasing the draft when a thick section of sliver enters the frame and decreasing it for thin sections. On the drawing frame, however, there is inevitably a slight time-lag between the time of measuring the sliver thickness at the back of the frame and the proper time for drafting that particular piece of sliver. In order to store the weight profile of the sliver a 'memory' is required which collects the information from the detecting rollers about the variations in sliver count, stores this information for a certain time, and then transmits it to the variable speed motor so that the latter can act at the proper time. B.J.T.R.A. hold British Patent 889,969 for such a device. By using such a machine considerable improvements can be made on the long-term regularity of the material but, inevitably, the machine is more costly than conventional fixed-draft frames.

CHAPTER EIGHT

Roving

For hessian and sacking qualities the roving frame has been superseded by the finisher drawing frame with its crimped sliver, but it is still used to produce heavy count 'rove' yarns in the range 1–7 ktex (70 to 200 lb/sp) or to provide another drawing stage to reduce the sliver count to a level suitable for spinning fine yarns of 120–170 tex ($3\frac{1}{2}$–5 lb/sp).

The roving frame is essentially a drawing frame fitted with an attachment for inserting twist into the drafted strand and winding it up on to a bobbin. The amount of twist that is put in depends upon whether the rove is to be used as a rove yarn or as a pre-spinning rove. For rove yarn sufficient twist must be inserted to give strength to the structure, but for pre-spinning roves only enough twist is put in to hold the fibres together to allow the material to be handled and to give some inter-fibre friction as it is being drafted on the spinning frame (see Chapter 9). Figure 8.1 illustrates a jute roving frame, with its gill-pins, positively driven flyer, and bobbin. There are three principal motions on a roving frame.

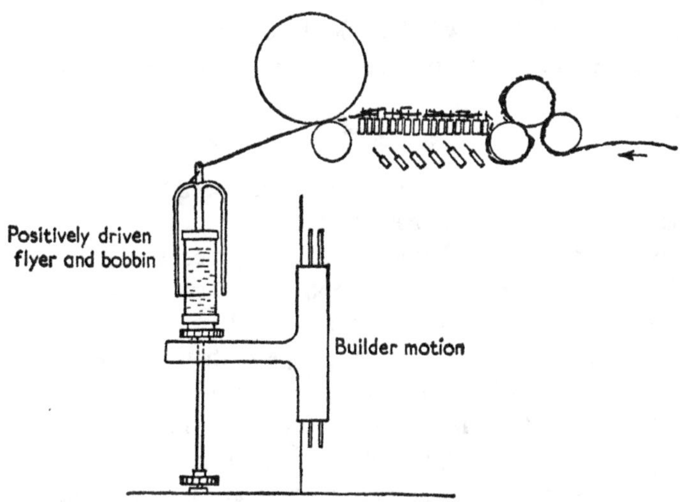

Figure 8.1. Essential features of the roving frame

(1) *Drafting.*

Drafting is carried out by the usual arrangement of retaining rollers and drawing rollers, with fibre control being exercised by gill-pins carried on faller-bars that are screw-driven. The factors governing the movement of the floating fibres that were discussed in the previous Chapter are also applicable here.

(2) *Twisting.*

The thin tenuous sliver emerges from the nip of the drafting rollers and passes down to the top of the flyer. It enters one of the hollow legs and travels down inside, to emerge near the foot and pass through the flyer 'eye'. As the flyer rotates, one end of the drafted strand is turned about the strand axis and the fibres become twisted into rove. The amount of twist which is inserted is changed when the count of the rove is altered (the reason for this will be dealt with later) and therefore some means must be found to do this. On the roving frame, the flyers are driven at a constant speed and so the only way to alter the amount of twist in the rove is to alter the speed of the delivery; if a low twist is desired then the material must issue from the drafting nip quickly, but if a high twist is wanted then the delivery speed must be reduced. For example, if the flyers rotate at 800 r.p.m. and 800 in. of rove are delivered each minute then there will be $800 \div 800 = 1$ turn of twist in each inch of rove, but if the delivery is reduced to 400 in./min then there will be $800 \div 400 = 2$ turns per inch. The relationship between flyer speed, twist, and delivery speed is,

$$t = \frac{n}{v}$$

where t is the twist per unit length, n is the speed of the flyers, and v is the delivery speed of the machine.

This is an important relationship since it means that the delivery speed of the machine is inversely proportional to the twist in the rove; thus a high twist automatically means a low delivery rate.

Since the flyers rotate at a constant speed they can be driven by a train of gear-wheels in the manner shown in Figure 8.2, the motion being derived from the main shaft. Because it is necessary at times to alter the twist by speeding-up or slowing-down the delivery rollers, the drive to these rollers is through a gear-train with a change pinion, the twist pinion, in it. When the twist pinion is changed the speed of the drawing rollers *and* the retaining rollers is altered but their relative speeds, i.e. the draft, remains unchanged.

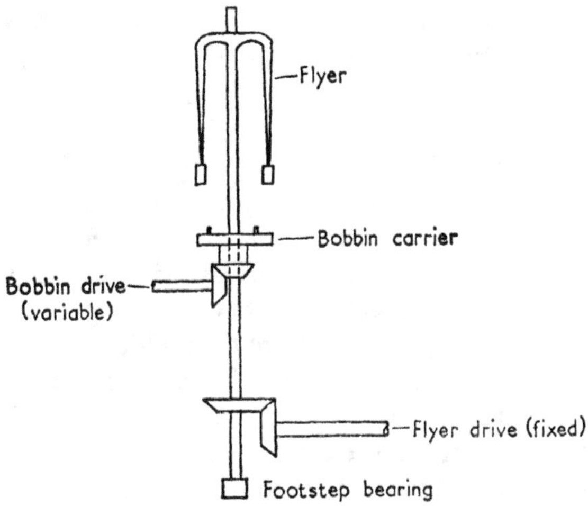

Figure 8.2. Flyer drive on a roving frame

(3) *Winding-on.*

At all times the delivery of rove from the drafting rollers must be wound up on to the bobbin. This is achieved by driving the bobbins slightly slower than the flyers, i.e. there is a flyer lead. (In other branches of the textile industry, bobbin lead may be found but as all jute frames are flyer lead only this type will be considered here.) The winding-on revolutions are equal to the difference between the flyer and bobbin revolutions, e.g. if the flyers rotate 700 times in a minute and the bobbins 600 times in a minute then there are 100 winding-on revolutions in a minute and the net effect is the same as if the bobbin has been stationary and the flyer has rotated round it 100 times.

On the roving frame, if v is delivery speed, d is bobbin diameter, n is winding-on revolutions, f is flyer revolutions, and b is bobbin revolutions, then,

$$n = (f-b)$$
$$v = \pi nd$$
$$v = \pi d(f-b)$$

As the delivery speed is fixed by the twist pinion on the frame and the flyer r.p.m. is fixed by the gearing, it follows from the above equation that as the diameter of the bobbin increases the winding-on r.p.m. must fall and, to accomplish this, the bobbin r.p.m. must *increase* as the bobbin fills.

In order to put as much rove on the bobbin as possible and to build a uniform package the coils of rove on the bobbin should lie neatly one above the other in a close-fitting spiral formation. This is achieved by mounting the bobbins on a movable carriage which can rise and fall and in so doing lift the bobbin into and drop it out of the flyers. This carriage is called the builder, and in the time taken to lay one coil of rove around the bobbin core it must move vertically a distance equal to the diameter of the rove if the rove is to fit snugly beside its fellow. At the start of the bobbin the circumference is small and one coil is put around quickly and therefore the builder must move equally quickly, but when the bobbin is nearly full then it takes longer to lay on a coil of rove and the builder must slow down to accommodate the increased laying time if the coils are to be laid contiguously.

The requirements of the winding-on motion can be summarized:

(1) It must increase the speed of the bobbins as they fill up.
(2) It must slow the builder down as the bobbins fill.

To accomplish this, a selection of mechanical devices may be used, such as expansion pulleys, friction plates, etc., but only one will be dealt with in detail here. This is the Holdsworth differential gear on the cone roving frame. The differential consists of a fixed bevel keyed to the main driving shaft, two free bevels carried in a straight spur gear wheel called the crown wheel, and a fourth bevel called the socket bevel which is attached to a free-running shaft *over* the main shaft. Figure 8.3 illustrates the device, with the socket bevel shaded. The crown wheel is positively driven in the *same* direction as the main

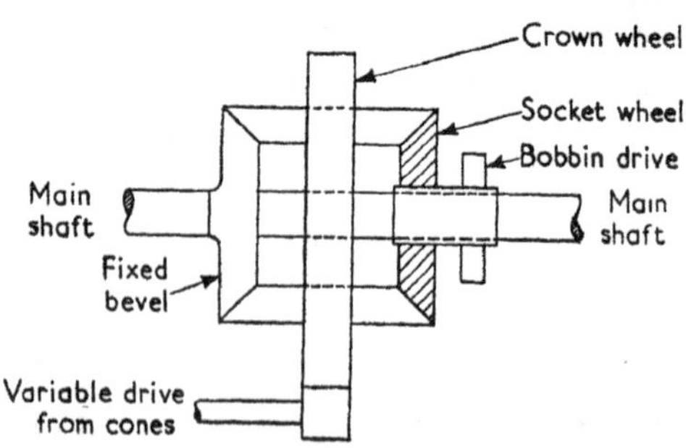

Figure 8.3. Differential motion on a roving frame

shaft. The socket bevel provides the drive to the bobbins and therefore if the speed of the bobbins is to be changed then the speed of the socket bevel must be varied first of all. Consider first the case where the crown wheel does not revolve; the fixed bevel runs at the main shaft speed, say 300 r.p.m., and through the bevels on the crown wheel acting as intermediates the socket bevel will be driven at the same speed, 300 r.p.m., but in the opposite direction as the fixed bevel. If the crown wheel is driven, each revolution makes the socket bevel rotate twice, i.e. if the crown wheel makes 30 r.p.m. the socket bevel will make 60 r.p.m. and so on. The only point still to be considered is the direction of rotation of the crown wheel. On jute flyer lead frames, the crown wheel always rotates in the same direction as the main shaft and each revolution of the crown wheel decreases the speed of the socket bevel by two revolutions. If the main shaft is running at 300 r.p.m. in a clockwise direction then, through the free bevels on the crown wheel, the socket bevel will run at 300 r.p.m. anticlockwise but, in addition, the crown wheel may be running at, say 30 r.p.m., in a clockwise direction like the main shaft. This clockwise motion drives the socket bevel at 60 r.p.m. also in a clockwise direction. The sum of the socket bevel r.p.m. then is 300 anticlockwise and 60 clockwise = 240 anticlockwise. The general form is

main-shaft r.p.m. − (crown wheel r.p.m. × 2) = socket bevel r.p.m.

Here, then, is a means of changing the speed of the bobbins during the time taken to fill one bobbin with rove; all that must be done is to *decrease* the r.p.m. of the crown wheel and the speed of the socket drive to the bobbins will automatically *increase*.

The general lay-out of the gearing of the cone roving frame is shown in Figure 8.4. The differential has been discussed already and it is the variable drive to the crown wheel which will now be dealt with. On this type of roving there are two cones—a top cone and a bottom cone—whose outlines follow a particular kind of curve called a hyperbola. The two cones are shaped in this way so that their combined diameters at any point is constant and the special speed considerations for the bobbin drive and the builder drive may be obtained. The top cone is driven through spur wheels from the main shaft at a constant speed and it, in turn, drives the bottom cone through a leather belt. Because of the shape of the cones the speed of the bottom one will vary depending on the position of the belt. For example, when the belt is at the extreme left-hand end of the cones where the top cone diameter

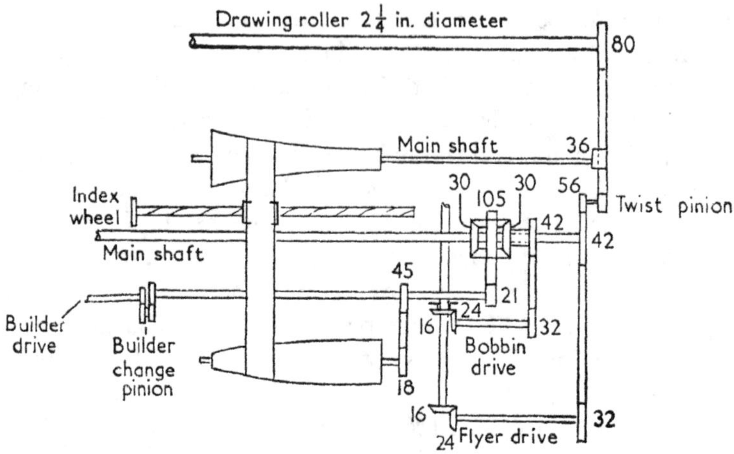

Figure 8.4. Cone roving gear

is 7 in. and the bottom cone diameter 3 in. then if the top cone is running at 240 r.p.m. the bottom cone will run at

$$240 \times \frac{7}{3} = 560 \text{ r.p.m.}$$

At the middle of the cones, the top diameter might be 5 in. and the bottom one 5 in., in which case the bottom cone would rotate at 240 r.p.m., but when the belt is at the right-hand end of the cones where the top cone diameter might be 4 in. and the bottom one 6 in. then the speed of the bottom cone would be 160 r.p.m. Therefore, by moving the belt along the cones the bottom one of the pair can be made to alter its speed. This speed variation is transmitted to the crown wheel of the differential through a train of gears. In this way the necessary alterations in speed of the bobbin drive take place.

Besides the change in bobbin speed to bring about the necessary winding-on, the builder speed requires to change to accommodate the different times taken to wind on one coil of rove on an empty bobbin and a full one. As can be seen from Figure 8.4, the builder is driven from the bottom cone through a train of gears and, therefore, as the speed of the bottom cone falls the builder slows down and allows more time for each coil of rove to be laid on the bobbin. There is a pinion in the gear train driving the builder called the traverse pinion which may be changed to give a general increase or decrease in the builder speed to suit different counts of rove.

It is now necessary to examine the way in which the belt is moved along the cones to effect the speed changes. Because the diameter of the bobbin increases by an amount equal to the rove diameter × 2 as each layer of rove is laid on, the speed of the bobbins (and the builder) should be changed at the end of each builder traverse. In other words, the speed of the bottom cones should not alter continuously but in a step-wise manner. This can be done by making the belt move along in regular steps at the end of each traverse of the builder. A simple mechanism, worked from the builder itself, is responsible for this.

As the builder comes to the top or bottom of its traverse it trips a small lever which allows a coarsely pitched pinion, called the index wheel, to move half a tooth. The index wheel is attached to a shaft which has a spiral groove cut into it along which runs the fork for moving the leather belt between the cones. At the start of each bobbin the belt is at the left-hand end of the cones and as the frame is started and the builder makes one traverse the index wheel is moved round half a tooth; this makes the shaft it is fixed to rotate slightly and the belt fork is moved along by the spiral. As time goes by, the belt is moved along the cones and the necessary speed changes are effected. When the bobbin is full the frame is stopped and the belt is pulled back by a hand-wheel ready for the start of the next bobbin. The rate at which the belt moves along the cones depends on the number of teeth in the index wheel. The index wheel must be changed to suit different sizes of rove.

ROVE TWIST

So far, only the mechanics of the roving frame have been examined, but it is now necessary to discuss rove twist in greater detail. When the fine ribbon of fibres is twisted together the fibres take up a spiral formation and the rove becomes more or less circular in section. The degree of twist can be expressed in two ways; in terms of the number of complete turns in a given length, or in terms of the angle at which the fibres are inclined to the axis. Figure 8.5(a) shows two roves, one much thicker than the other, having the same twist angle; if, however, these are examined from the point of view of the turns in a given length it will be found that they do not have the same number of turns. Twist angle and turns per unit length are related; in Figure 8.5 (b) the rove has been cut along its axis and opened out in one plane. It will be seen that a triangle is formed whose base equals the circumference

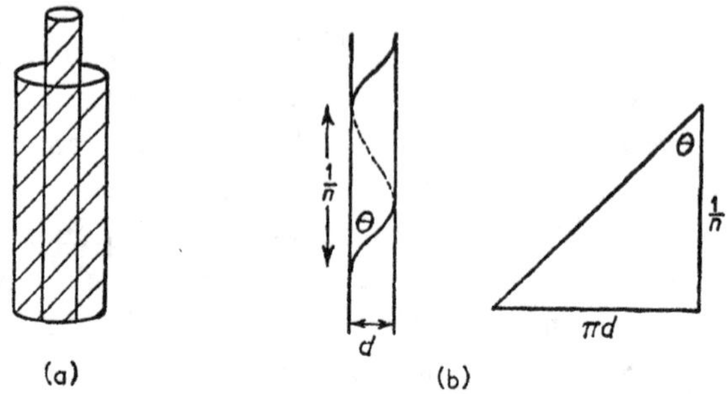

Figure 8.5. Twist and twist angle. n *= turns per unit length,* d *= rove diameter,* θ *= twist angle*

of the rove and whose height depends on the turns per unit length, or rather its reciprocal, the length of one turn. From the triangle,

$$\tan \theta = \pi d n$$

For practical reasons, it is easier to measure the turns per inch (or per centimetre) so twist is always referred to in these terms, but in fact it is the twist angle which is the important factor in deciding how the rove will behave. In a twisted structure, be it rove or yarn, if a tensile force is applied along the axis, the fibres, because of their angle, exert an inward-directed force which has the effect of increasing inter-fibre friction and making it more difficult for the tensile force to rupture the structure. If one has twistless rove, there is no inter-fibre cohesion at all and the fibres slip past each other as soon as a tensile force is applied, but if twist is inserted and steadily increased the strength of the rove gradually rises as more and more inter-fibre friction is induced by the inward-directed force resulting from the spirality of the fibres. After a certain point, however, any increase in the twist angle cannot compact the yarn any further and the fibres are now in a state of strain and the strength of the rove begins to decrease. Thus, if the strength of the rove is plotted against the twist angle, as in Figure 8.6, one sees a steady increase over the part (a), a flat maximum over (b), and a fall in strength over (c). In (a) the rove breaks by the fibres slipping past one another, in (b) there is a mixture of fibre-slip and fibre-breakage, and in (c) the predominant cause of failure is fibre-breakage because the inward force is sufficiently strong to stop fibre-slip. The use to which the rove is to be put, therefore, determines how much twist will be inserted. If it is to be used as a pre-spinning rove, where the fibres must be able to slip past

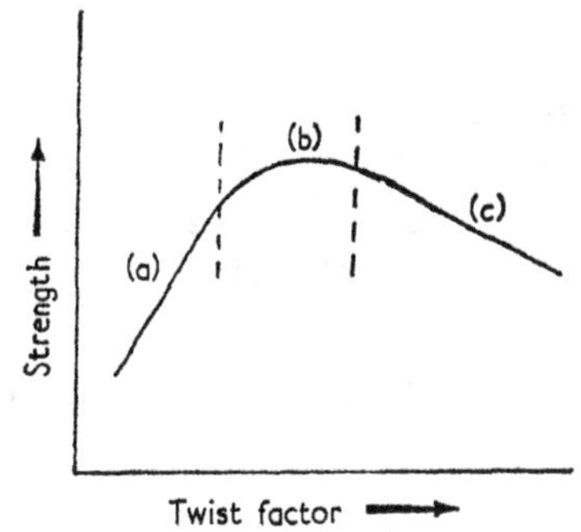

Figure 8.6. Twist/strength relationships

one another during drafting on the spinning frame then, obviously, one must work on the (a) part of the curve, but if the rove is to be used as a heavy count yarn where strength is required then the twist must be selected which would give a strength in the (b) part of the curve.

If one has roves of different count and wishes these roves to have the same *relative* degree of strength, then one must arrange for the twist angle to be the same in each case, so that the inward-directed forces will be equal. The twist angle is related to n, the turns per unit length, and d, the rove diameter, but d is proportional to $\sqrt{(\text{count})}$, therefore, θ is related to $\sqrt{(\text{count})}$.

$$\tan\theta = \pi d n$$

$$d \propto \sqrt{(\text{count})}$$

$$\tan\theta \propto n\sqrt{(\text{count})}$$

$$\theta \propto n\sqrt{(\text{count})}$$

i.e. if θ is to be constant for all weights of rove

$$n\sqrt{(\text{count})} = K$$

where K is a constant, known as the twist factor. By means of this twist factor it is a simple matter to calculate the turns per unit length for any count of rove. In jute units the twist factor for pre-spinning rove is normally about 7 and for rove yarns about 10.

$$\text{turns per inch} = \frac{7}{\sqrt{(\text{lb/sp})}} \quad \text{for pre-spinning rove}$$

$$\text{turns per inch} = \frac{10}{\sqrt{(\text{lb/sp})}} \quad \text{for rove yarns}$$

In the tex system,

$$\text{turns per centimetre} = \frac{16}{\sqrt{(\text{tex})}} \quad \text{for pre-spinning rove}$$

$$\text{turns per centimetre} = \frac{23}{\sqrt{(\text{tex})}} \quad \text{for rove yarns}$$

PINION CHANGES NECESSARY WHEN CHANGING COUNT

(1) *Draft.* This pinion is arrived at in the usual way,

$$\text{draft pinion} = \frac{\text{draft constant}}{\text{draft}}$$

$$= \frac{\text{draft constant} \times \text{rove count}}{\text{sliver count}}$$

(2) *Twist.* The required pinion can be found from a gearing constant called the twist constant which is analogous to the draft constant. It is found by assuming that the drawing roller is driving the flyers and calculating the number of flyer revolutions made in the time taken for one revolution of the drawing roller. This number of turns of the flyers is then inserted into the length of rove delivered by one revolution of the drawing roller, i.e. one circumference. The pinion is found from

$$\text{twist pinion} = \frac{\text{twist constant}}{\text{turns per unit length}}$$

(3) *Index.* The speed of the bobbins (depending on the bottom cone speed and the crown wheel speed) must be changed in proportion to the diameter of the rove since the bobbin diameter is increased by twice the rove diameter as each layer is put on. It is more convenient to work in terms of count than diameter as the latter is directly proportional to the square root of the count. The index wheel chosen is not rigidly fixed like the draft and twist pinions as individual preference may decide just how tight the rove is to be wound on the bobbin, but it is common practice to work with an index constant of 135, i.e.

$$\text{index wheel} = \frac{\text{index constant}}{\sqrt{(\text{count})}}$$

(4) *Builder.* Similarly to the index wheel, the builder pinion is not absolutely fixed by the rove specifications but the pinion should be such that the builder is driven up and down at a speed that will lay the

coils of rove side by side. In this way as much rove as possible will be packed on to the bobbin. Again, a builder constant may be used, 280 being a common one

$$\text{builder pinion} = \frac{\text{builder constant}}{\sqrt{(\text{count})}}$$

Table 8.1 summarizes these changes.

TABLE 8.1

Pinion	*Heavy rove*	*Light rove*	*Pinion Proportional to*
Draft	More teeth	Fewer teeth	Count
Twist	More teeth	Fewer teeth	√(Count)
Index	Fewer teeth	More teeth	√(Count)
Builder	More teeth	Fewer teeth	√(Count)

PRODUCTION ASPECTS OF THE ROVING FRAME

The common sizes of roving frame range from 56 to 80 spindles with a production capacity of 300–400 lb/hr. The efficiency (running time ÷ total time) of roving frames is usually around 70–80 per cent, much of the lost time being due to doffing. To doff the full bobbins of rove each flyer must be given a half-turn and lifted off its spindle, the bobbin of rove removed and an empty one substituted, and then the flyer replaced. This, as may be imagined, takes some little time.

The rove bobbins in common use are 10 in. long by 5 in. or 6 in. in diameter and work with a packing density of about 26 lb/ft^3. The following example is typical of those met.

A rove bobbin holds 2·9 lb of material. A 64 lb/sp pre-spinning rove is being produced at a flyer speed of 600 r.p.m. If doffing takes 3 min, what is the machine efficiency, allowing 10 per cent for unavoidable stoppages due to mechanical troubles, etc.?

The bobbin holds $= \dfrac{2{\cdot}9 \times 14400}{64}$

$= 645$ yd of rove

Twist in rove $= \dfrac{7}{\sqrt{64}}$

$= 0{\cdot}88$ t.p.i.

Delivery speed of frame $= \dfrac{600}{0{\cdot}88 \times 36}$

$= 19$ yd/min

Time to fill the bobbin $= \dfrac{645}{19}$

$= 33$ min

Total cycle time (including doffing) $= 36$ min

Machine efficiency $= \dfrac{33}{36} \times 100 - 10$ per cent

$= 82$ per cent

CHAPTER NINE

Spinning

The majority of jute yarns are spun from finisher drawing sliver and spinning from rove is confined chiefly to the finer counts of yarn (173 tex, 5 lb/sp or less). The advantage of using crimped sliver is an economic one, for the cans of finisher drawing sliver hold sufficient material for 25–30 hr spinning compared with about 5 hr supply on a bobbin of rove. As a result of this increased package size, less labour is required for material handling. The move towards sliver spinning has been accompanied by the use of longer drafts at the spinning frame with the accompanying reduction in the number of deliveries required to supply the spinning frames.

The essential features of the spinning process are drafting, twisting, and winding-on. Spinning frames are made in several different sizes, designated by the distance between adjacent spindles, i.e. the pitch. Only a small part of the entire count range is produced on a given pitch of frame but, no matter what the size of the frame, the mechanisms for twisting and winding-on function in the same manner although some differences exist in the methods adopted for controlling fibre motion during drafting.

DRAFTING

All jute spinning frames have two sets of rollers extending along the whole length of the machine—the retaining rollers and the drawing rollers. Each of these sets consist of a positively driven member and a pressing member, between which the fibres are gripped. The draft operates in the usual way by attenuating the material and reducing its count.

The different types of spinning frames can be classified according to their method of draft control.

(1) Breast plate.
(2) Breast plate and intermediate rollers.
(3) Apron and intermediate roller.
(4) Double apron.

(5) Grooved intermediate rollers.
(6) Gill-pins.

The first type is confined to rove-spinning and the remainder to sliver spinning. The drafting mechanisms of the various types are illustrated in Figure 9.1.

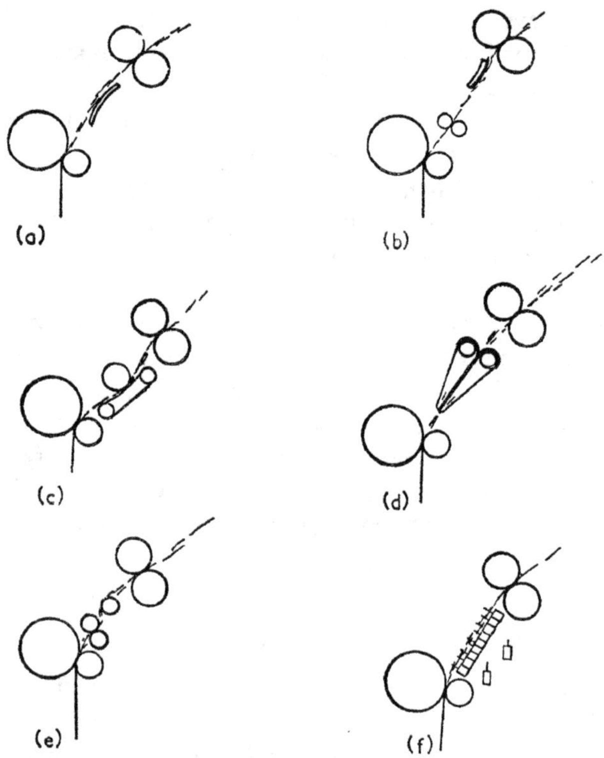

Figure 9.1. Methods of fibre control on jute spinning frames

(1) *Breast plate* (*Figure 9.1*(*a*))

The reach of the frame is, as usual, slightly longer than the length of the longest fibres in the material. Situated between the retaining and drawing rollers is a smooth metal plate called the breast plate. This plate projects forward slightly from the line joining the nips of the two sets of rollers in order that it may play its proper part in the control of short fibre movement. The twist in the roving, it will be recalled, is of such a magnitude that some degree of cohesion is imparted to the strand but at the same time inter-fibre movement is not impeded. When the rove passes down the breast plate towards the drafting nip its leading fibres are caught between the drafting and pressing rollers and pulled from the rove. This has the effect of reducing the count of the rove and, in so doing, the twist angle becomes progressively less and less in relation to the rove count, e.g. the rove may enter the drafting field weighing 84 lb/sp and having 0·75 t.p.i. but by the time a few fibres have been drafted from it the count may only be 50 lb/sp, reducing the twist factor from 6·9 to 5·3. Consequently the inward-directed forces arising from the twist are less. Under these circumstances less restraint is applied to the fibres and hence some degree of draft control is lost. The function of the breast plate is to determine where the rove will begin drafting. This it does by virtue of its position. Because of drafting a slight tension develops in the rove which presses the material more firmly on to the plate. The tension below the plate is greater than that on or above the plate and since inter-fibre movement occurs at the point subjected to the greatest tension, it is here that drafting takes place. By altering the position of the plate relative to both sets of rollers the tension in the rove can be increased or decreased; an increase restricting drafting until just before the drawing nip, a decrease allowing earlier drafting. The draft control from the combination of breast plate position and rove twist is not of a high order and the setting of the plate does not appear to be critical. It is customary to site it in such a manner that the rove is just beginning to untwist as it approaches the foot of the plate.

Immediately beneath the breast plate there is a small conductor for leading the fibres right into the drafting nip. It, too, can be adjusted inwards and outwards.

(2) *Breast plate and intermediate rollers*

This is one of the commonest methods adopted for draft control in jute spinning at the present time. The frame is designed for use with crimped finisher drawing sliver and is illustrated in Figure 9.1(b). In many ways, its design and operation are similar to those of the type just described for rove spinning with the exception, of course, that the material enters the drafting field without any twist in it. The breast plate in this case is a small semicircular plate, concave outwards, which can be swung on its own axis and moved bodily inwards or outwards. The sliver passes down behind the plate and then enters a short channel at the foot of which there is a pair of intermediate rollers, the lower one being positively driven and the upper deriving its motion from the lower of the pair. Both rollers are deeply fluted, the upper having a groove cut in its surface to allow the sliver to pass through. The upper roller is self-weighted and as the sliver passes underneath a gentle restraining force is applied, insufficient to stop drafting but great enough to prevent much premature drafting of the short fibres. After leaving this pair of rollers the sliver enters a small conductor and then passes directly into the drafting nip. On this type of frame the siting of the various members sets up a tension in the sliver when it is being drafted; a tension which causes the material to bear more heavily on the breast plate and consequently increases its resistance to short fibre movement. Thus drafting, with the exception of a small amount of long fibre movement, does not take place until the sliver is between the nip of the intermediate rollers.

(3) *Apron and intermediate roller*

In this type of draft control, Figure 9.1(c), the breast plate has been discarded in favour of an endless rubber apron. The fibres leave the nip of the retaining rollers and then pass on to the surface of a rubber apron. As they move down this, they meet an intermediate roller which is pressing gently into the apron—this helps to stop uncontrolled fibre movement. Below the apron is the usual conductor just before the drafting nip.

(4) *Double apron (Figure 9.1 (d))*

This type is a more recent development of the one just described, in which the intermediate roller has been replaced by a second rubber apron. The sliver passes down between the aprons and the fibres are gripped continuously. The lower apron is driven by a grooved wheel at its upper end, and its lower end is made to turn sharply round a small adjustable plate. The upper apron is driven by contact with the lower and similarly passes round a small plate at its lower end. In this way both aprons can be brought very close to the drawing nip and a positive grip maintained on the short fibres as late as possible.

(5) *Grooved intermediate rollers*

This type of control, Figure 9.1(e), is confined to some large pitch frames used for heavy yarns. The sliver passes down over a series of smooth-surfaced intermediate rollers, each of which has a deep groove cut in its face. The siting of the lower rollers can be adjusted to give a greater or lesser tension in the sliver. In the same manner as type (2), the upper members of the intermediate pairs are self-weighted.

(6) *Gill-pins*

The gill-spinning frame is very similar to the spiral gill-drawing frames described earlier, with the difference that the faller bed is inclined at approximately 45 degrees. To suit the nature of the material at this stage in the process the gill-pins are fine and densely set. The spirals are usually $\frac{1}{4}$ in. pitch and triple screw. This type of frame is limited in its speed capabilities by the faller drops per minute just as the drawing frames are, about 500 per minute being considered the maximum. However with the speed limitations imposed by the flyer design and the quality of the yarn (discussed later) it is seldom that the frame works at the maximum faller-bar speed.

These, then, are the types of draft control found on jute spinning frames. As the variation in the count of short lengths of yarn (the 'thicks and thins') is largely decided by the regularity of the finisher drawing sliver and the manner in which the spinning draft is applied, it is desirable that the draft control mechanism should operate as efficiently as possible. Spinning draft is changed by means of a change pinion and, in the usual manner for jute machinery, when the draft is altered it is the feed speed which changes, the delivery speed remaining constant. Indeed, on the spinning frame this is essential, for any change in the front roller speed causes a change in the twist in the yarn. It is at the spinning frame that draft changes are made to produce yarns to suit sales requirements and therefore it is essential that the correct draft be selected. The draft imposed upon the material must be such that the yarn is spun to the correct count; for this reason a careful assessment must be made of two factors which affect yarn count. These are moisture regain and twist take-up.

Finisher drawing sliver usually has a moisture regain of about 26 per cent and from this material a yarn must be spun which will have the correct count when it is dispatched. During spinning some 25 per cent of the sliver moisture will be evaporated and during the subsequent processes of winding and in storage a further 15–25 per cent will be lost. The yarn must be taken off the frame at a count slightly above the required level to allow for these post-spinning moisture losses. For this reason the draft must be *reduced* by a certain amount. The exact amount of the decrease depends upon the moisture level before, during, and after spinning, but it is customary to arrange for the yarn to have the correct count at 14 per cent moisture regain.

Twist take-up will be dealt with more fully later, suffice it at this stage to say that take-up increases the yarn count and therefore the draft must be *increased* to allow for this. The amount of take-up depends on the degree of twist but for normal twist factors it is between 2 and $2\frac{1}{2}$ per cent.

The method of calculating the required spinning draft is as follows

$$\frac{S}{Y}\times\frac{100+R_y+O}{100+R_s+O}\times\frac{100+T}{100}=D$$

where S is the sliver count, Y the yarn count at R_y per cent regain, R_y the yarn regain at which the yarn will be of the correct count (per cent), R_s the finisher drawing sliver regain (per cent), O the oil content (per cent) (on dry fibre basis), T the twist take-up (per cent), D the spinning draft.

For example, 150 lb/sp finisher drawing sliver, with a regain of 25 per cent is to be spun into 8 lb/sp yarn which will have the correct count at 14 per cent regain. The twist take-up is 2 per cent, and the oil content is 6 per cent. What spinning draft is required? What will the yarn count at the frame be if the regain of newly spun yarn is 19 per cent?

Spinning draft

$$\frac{150}{8}\times\frac{120}{131}\times\frac{102}{100}=17{\cdot}5$$

Count at 19 per cent regain

$$\frac{150}{17{\cdot}5}\times\frac{125}{131}\times\frac{102}{100}=8{\cdot}34 \text{ lb/sp}$$

It is customary to check the count of the yarn at the spinning frames as this is the last point where corrective action can be taken if required. Testing is done by taking hanks off a number of bobbins and weighing. For jute yarn testing, the standard reel for winding test-hanks is 90 in. in circumference, 40 turns of the reel making 100 yd. It sometimes happens that over a period of time the count drifts up or down, but unless one is sure that this drift is genuinely due to a change in the *fibre* content of the yarn, no draft pinion change should be made. If, as may happen, such a drift is due to moisture regain changes then a draft change would lead to the wrong count of yarn being spun For this reason the moisture regain should always be checked when a count test is made; more will be said about this in a later chapter, but an example may help to show how this occurs.

Suppose that the yarn in the previous example is being spun, but because of an unusually low relative humidity in the spinning department, the yarn regain falls to 16 per cent at the spinning frame.

New count at frame

$$\frac{150}{17{\cdot}5}\times\frac{122}{131}\times\frac{102}{100} = 8{\cdot}14 \text{ lb/sp}$$

If the draft pinion in use was a 36 tooth, then in order to bring the yarn back on count a pinion change might be made.

New pinion required

$$\frac{8{\cdot}14}{8{\cdot}34}\times 36 = 35 \text{ tooth}$$

The new draft with a 35 tooth pinion would be

$$17{\cdot}5\times\frac{35}{36} = 17{\cdot}0$$

The new yarn count at 14 per cent regain would now be

$$\frac{150}{17{\cdot}0}\times\frac{120}{131}\times\frac{102}{100} = 8{\cdot}24 \text{ lb/sp}$$

Clearly, in this example the yarn count has been made 'off standard' because of a wrong decision. A draft change was made because of an alteration in the yarn moisture.

TWISTING

Jute spinning frames insert the twist by means of overhung flyers suspended above the bobbins. There is no positive drive to the bobbins as there is on the roving frame and the bobbins are made to rotate by the yarn pulling them round. Figure 9.2 shows the twisting arrangement adopted. The flyers are carried on ball-bearing wharves mounted on the front of the frame at about waist-height. The part of the wharf projecting above the mounting assembly is called the 'cap' and plays an important part in the actual operation of the frame, as will be seen later. The wharf is driven by a cotton or nylon tape from the main cylinder of the machine, that part where the tape runs being crowned

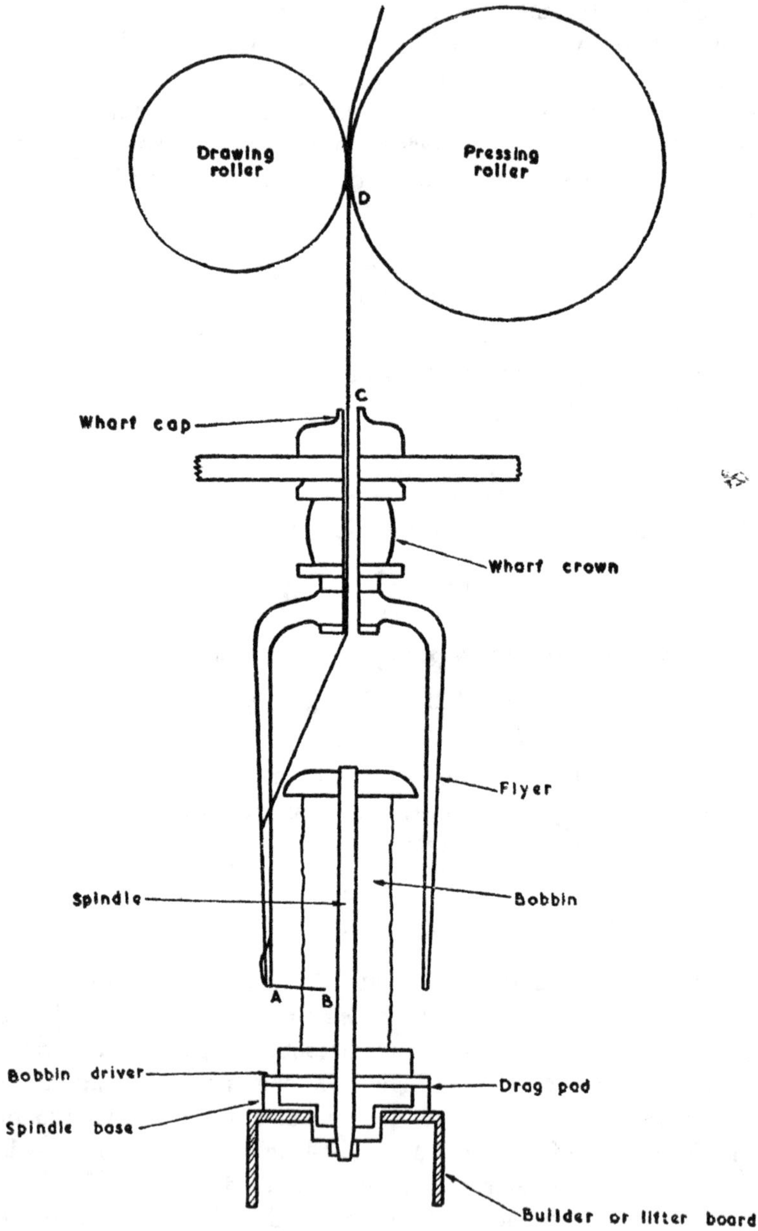

Figure 9.2. Twisting and winding-on section of a jute spinning frame

so that the tape does not give an erratic drive by wandering up and down the bearing surface.

The yarn passes down from the drafting nip to the top of the wharf cap where it enters a central hole and continues down through the wharf. At the exit of the hole a ceramic disk is cemented to protect the metal from the abrasive action of the yarn. The flyer legs are screwed on to the wharf so that they may be replaced if necessary. The legs themselves are tapered towards their tips to reduce centrifugal 'throw-out'. The flyer legs have a small 'eye' at the foot through which the yarn passes on to the bobbin. As the flyers are designed to run at high speeds they must be dynamically balanced otherwise any eccentricity would ultimately damage the whole assembly and could cause a serious accident.

The simplest relationship between flyer speed, delivery speed, and twist is

$$t = \frac{n}{v}$$

where t is the turns of twist per unit length, n the flyer speed, and v the delivery speed. This equation, however, must be modified in the light of twist take-up. If a ribbon of untwisted fibres is rotated about its own axis and twist inserted then it inevitably becomes shorter as the fibres assume a spiral formation. The amount by which the structure reduces in length is known as the 'take-up' and is expressed as a percentage. Thus

$$\frac{\text{untwisted length} - \text{twisted length} \times 100}{\text{untwisted length}}$$

The exact amount of take-up depends upon the twist angle in the yarn; the greater this angle the more take-up there is. Figure 9.3

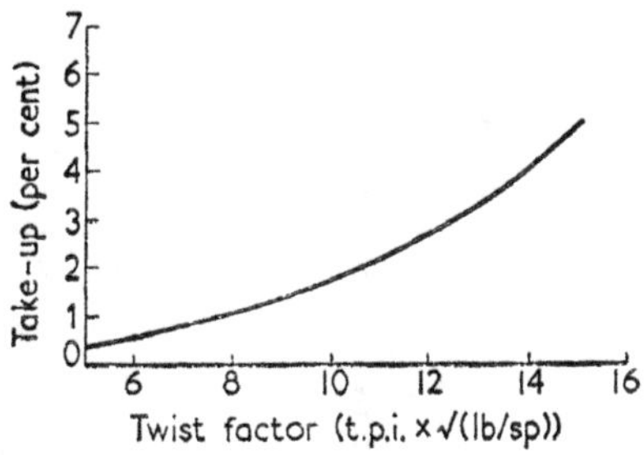

Figure 9.3. Effect of twist factor on twist take-up

shows the relationship between twist angle, as expressed by the twist factor, and take-up for jute yarns. It will be seen that for the common range of twist factors the take-up is of the order of 2 or $2\frac{1}{2}$ per cent. Just as the count is increased by take-up so the twist in the yarn is increased by take-up and therefore the equation above should be altered to

$$t = \frac{n\,(100+T)}{100\,v}$$

where t, n, and v have the same meanings as previously and T is the percentage take-up.

Even using this equation does not, however, give the full picture of yarn twist. If a yarn is examined closely it will be found that the number of turns of twist varies from point to point along the length. This arises chiefly from the fact that the yarn mass itself fluctuates from point to point. Yarn twist is inserted by rotating the lower end of the yarn about the upper end and the twist actually ascends from below into the upper portions of the yarn and in this way runs up towards the drawing nip. The twist is transmitted by the lower fibres taking up a spiral formation and forcing those above them to conform to the same configuration. The fewer and less rigid the fibres the easier is it for the lower ones to force the upper ones to take up the same twist angle as themselves. Notice again that it is the twist angle which is the same along the length of the yarn. Because the twist angle is constant (or in more practical terms, the twist factor is constant) those parts of the yarn that are thin have more turns per unit length than those that are thick.

Common twist factors in use are shown in Table 9.1.

TABLE 9.1

	lb/sp and turns/in.	*tex and turns/cm*
Sacking weft	10·0	23·0
Hessian weft	10·5	24·3
Hessian warp	11·5	26·2
Carpet yarns	12·5	29·0
Sacking warp	13·0	30·2

WINDING-ON

As the action of the builder is the simpler of the two winding motions it will be dealt with first. The bobbins rotate around central dead spindles which are set vertically in the builder. As the builder moves up and down the bobbins alternately rise into and withdraw from the flyers and this reciprocating movement, combined with the rotation of the flyers about the bobbins, winds the yarn on the bobbin in a continuous spiral. Notice that when the builder is at the top of its traverse the yarn is winding on at the bottom of the bobbin and *vice versa.*

The builder is suspended on short lengths of chain which are attached to pulleys keyed to a shaft running along the whole length of the frame. Brackets from the builder carry sleeves which run up and down on columns to give steadiness and stability to the motion. The traversing movement is obtained from a lever at one end of the frame which is made to rise and fall by a heart cam underneath it. Figure 9.4 illustrates the principal parts of the builder motion. The length of the traverse depends on the throw of the cam, the length of the lever following the cam, and the diameters of the pulleys marked A and B in the Figure. There are turnbuckles in the linkage connecting the lever arm to the pulley shaft so that the position of the builder

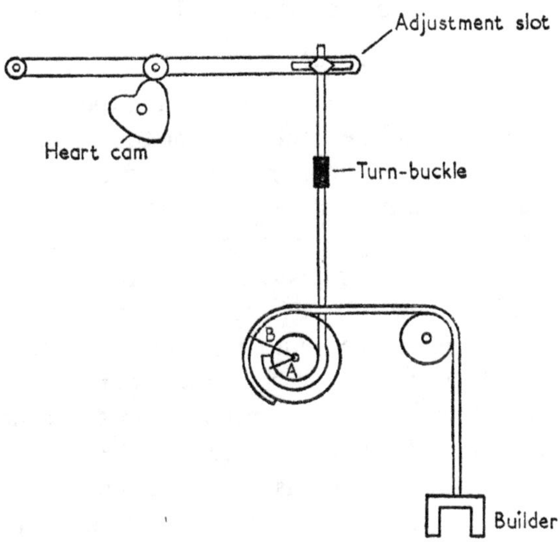

Figure 9.4. Spinning frame builder motion

relative to the flyers and bobbins can be adjusted. The builder should change direction just at that moment when the yarn is winding on at the flange of the bobbin. If the builder is too high or too low in relation to the bobbins then the yarn will be built up unevenly on the

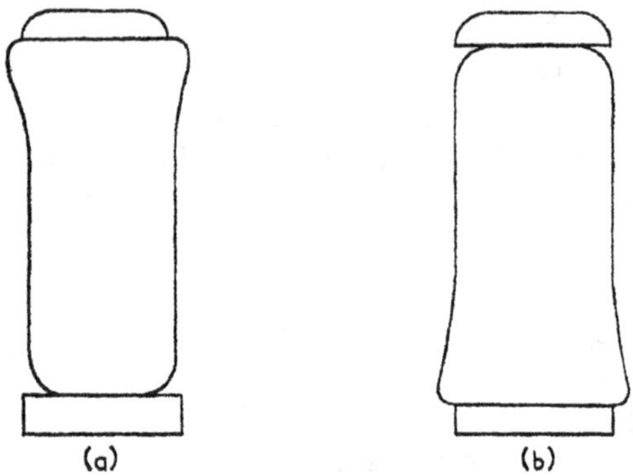

Figure 9.5. Bad bobbin building due to faulty positioning of the builder: (a) *Builder too low;* (b) *builder too high*

bobbin. Figure 9.5 shows the shape of the bobbin when the builder is not adjusted properly. Bobbin building like this is undesirable as it affects the spinning tension adversely, as will be seen later.

In Chapter 8 it was shown that in order to achieve correct winding-on the following equation had to be fulfilled

$$(n-b)\,\pi d = v$$

where n is the flyer r.p.m., b the bobbin r.p.m., d the bobbin diameter, and v the delivery speed. The implication of this relationship is that as the bobbin fills, its revolution rate must increase in order that $\pi(n-b)$ will decrease as d becomes greater. On the roving frame the bobbins were driven through direct gearing at a speed which was varied by the cones and differential. But as the bobbins on a spinning frame are not driven how is this increase in bobbin revolution rate attained? In fact, it is attained automatically by the bobbins themselves as a result of the manner in which they are rotated by the yarn. Figure 9.2 shows how the bobbin rests on a metal carrier which is mounted on a 'dead' spindle on the builder. The upper part of the carrier is a hollow sleeve, which is a loose fit on the dead spindle, and the lower part a flange just slightly larger than the bobbin base. Two small pegs project from the flange of the carrier and when the bobbin is slipped on to the carrier these pegs fit into recesses cut into the underside of the bobbin base. By means of these pegs there is a loose but positive drive between the carrier and the bobbin and they rotate as a pair about the central dead spindle.

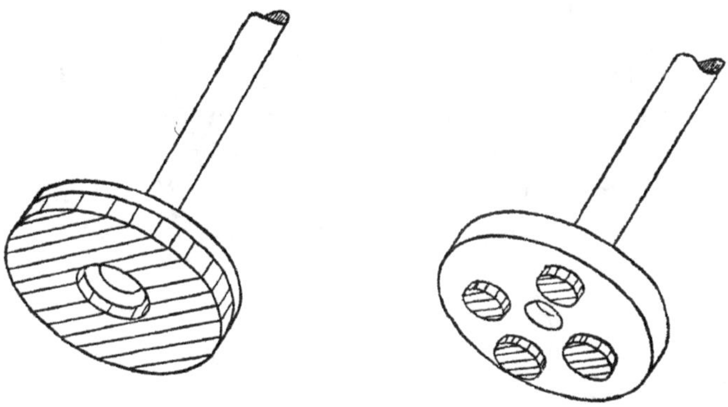

Figure 9.6. Bobbin carrier friction pads

Figure 9.6 shows the underside of two bobbin carriers. One has a complete ring of felt attached to it (shaded in the Figure) and the other has four small felt pads instead of the ring. Whether the solid ring or the pads are used the principle of operation is the same. When the carrier is in position on the builder the felt pads bear against a smooth plate encircling the dead spindle and when the carrier is rotated these felts set up a drag by virtue of the friction between them and the bearing plate on the builder. During the spin it is the yarn which pulls the bobbin and the carrier round and two equal opposite forces act—a tension in the yarn pulling the bobbin round and a drag tending to prevent the bobbin moving. Because of their function these felt pads are known as 'drag-pads'. These contra-acting forces are turning-forces or torques and their magnitude is found from the product of the force and its moment about the central point. For instance if the bobbin radius at some instant during the spin is 1·25 in. and the tension in the yarn turning the bobbin round is 0·9 lb, then the torque that is rotating the bobbin is

$$1{\cdot}25 \times 0{\cdot}9 = 1{\cdot}125 \text{ in.lb}$$

Similarly, if the frictional force of the drag-pads is assumed to be concentrated at the mid-point of the pads at a distance r from the centre of rotation and have a magnitude p, the torque opposing motion is rp.

While the bobbin is rotating, these two torques are equal. Hence,

$$rp = RT$$

where r and p have the meaning given above, R is the radius of the bobbin, and T is the tension in the yarn between the eye of the flyer and the surface of the bobbin.

In order that the bobbin keep rotating, energy must be supplied to the system. This energy comes from the flyer, but of course is ultimately derived from the frame motor. As the bobbin fills up, more energy is required to turn it since it not only is becoming heavier but is also rotating at a higher speed. The torque RT steadily increases throughout the spin but as the bobbin radius increases at a much

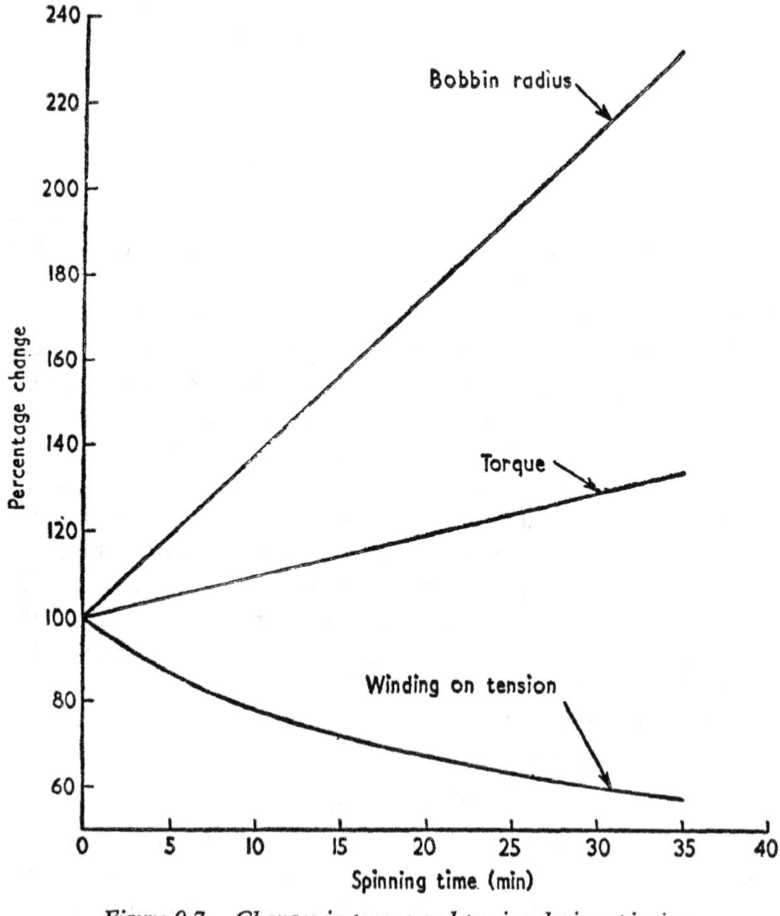

Figure 9.7. Changes in torque and tension during spinning

faster rate than does the torque, the yarn tension, T, becomes smaller as the bobbin builds. Figure 9.7 shows this effect; for convenience each of the variables has been expressed as a percentage, the starting values being taken as 100 per cent in all cases. It will be seen that although the torque required to keep the bobbin turning grows steadily, the yarn tension falls throughout the spin, always remembering that

$$\text{tension} = \frac{\text{torque}}{\text{radius}}$$

It was said above that before the bobbin will rotate continuously, energy must be supplied and that this was done through the flyer pulling the yarn round. If therefore the yarn breaks, then this supply of energy is immediately lost and the frictional torque of the drag-pads brings the carrier and the bobbin to a halt. The yarn will break if the tension in it is greater than the strength of the weakest point. Yarns break frequently, therefore, if either the yarn strength is low or the tension is high. Yarn strength is very largely a matter of the grade of fibre being used and so to avoid an excessive number of spinning breaks the tension should be as low as practicable. Spinning tension, like so many other factors in jute processing, is a balance between two extremes. The upper level of tension is determined by the ability of the yarn to spin successfully; if this level is exceeded then a large number of breaks will occur, obviously this level is related closely to yarn count and strength. The lower level is set by the phenomenon known as 'ballooning'. Ballooning occurs when the tension in the yarn is insufficient to hold it against the flyer leg and the centrifugal force of rotation throws the yarn off the leg in a wide balloon. When this happens the yarn strikes the adjacent flyer and breaks. Since the heavier the yarn the more tendency there is to ballooning, it is necessary to apply greater tension to keep the yarn on the flyer leg; fortunately the heavier yarns can withstand the greater axial tension that is required. On the four-pad type of carrier the pads are put in the outer position when heavy yarns are being spun for this very reason. On the other hand, when a light count is being produced they are placed in their inner position, so that the frictional torque is at its lowest value.

Frictional torque at the drag-pads depends upon such factors as the weight of the assembly, the clearance between the carrier sleeve and the spindle, etc., which are set by the machinery designer and, as such, are outside the control of the user. There are, however, two important features that the user can control—the amount of friction between the pads and the builder bearer plate, and the effective friction radius.

The friction radius can be taken, without serious error, as half-way between the inner and outer edges of the pad and, as in a simple lever, the greater this distance is from the centre of rotation the greater will be the frictional torque and, consequently, the higher the yarn tension. On the full-pad type the friction radius can be altered only by reducing the radius of the pad in contact with the builder; two methods are available for this. Firstly, a smaller drag-pad may be put on or, secondly, a smaller builder bearer plate may be substituted. This latter method is used by one machinery maker for varying the tension to suit the count of yarn being spun or, alternatively, to alter the tension while the frame is in motion, see Figure 9.8. The bearer plate is

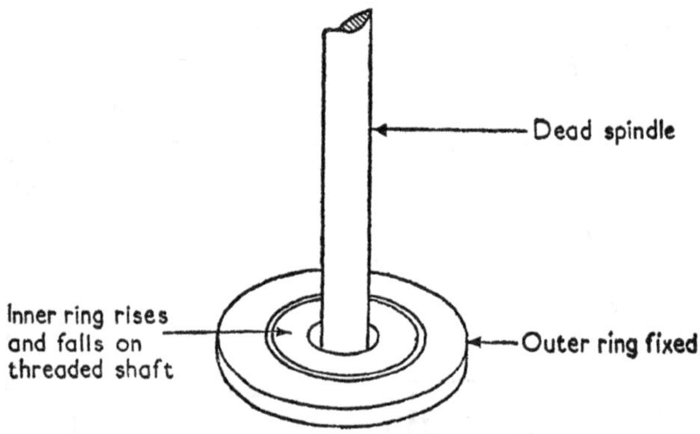

Figure 9.8. Two-position builder friction plate

made in two concentric parts, the inner one of which is mounted on a short threaded spindle. By means of a handle attached to the spindle and projecting from the front of the builder, the central ring may be raised or lowered at will. When it is raised it is just above the level of the outer ring and the felt drag-pad bears only on the inner ring. Consequently, the friction radius is small and the spinning tension low. When the inner ring is lowered the friction radius is greater and the frictional torque is increased. This method of operation allows the frictional torque to be varied throughout a spin and the system is used in an attempt to keep the spinning tension fairly level during the spin. It will be recalled that with a fixed frictional radius spinning tension was highest at the start of each new bobbin and fell gradually throughout the spin; if this two-ring system is used then it is possible to work with a lower frictional torque (and spinning tension) at the start of each bobbin, and as the bobbin diameter grows, the frictional torque can be increased by lowering the inner ring.

The four-pad type gives another method by which the frictional radius can be altered. It should be noted, however, that this cannot be done during the spin but only when the bobbins are not in use. Each small pad is mounted upon a short spring arm which can be put into one of three positions—inner, middle, or outer. When the pads are in their innermost position the frictional radius is small and consequently the spinning tension is low; when the pads are at the outer position, spinning tension is high. To suit the requirements of the yarn as far as ballooning and axial tension are concerned, the pads are placed at the inside position when light counts are being spun and at the outside when the heavy end of the count range for the frame is being produced.

Thus the manufacturer can, within the limits of the system on his frames, alter the general level of tension by means of these devices that alter the friction radius. There is another important factor over which he can exert some control. This is the amount of friction developed between the drag-pad and the bearing plate on the builder. Most drag-pads are made of felt, but for situations where an extra low tension is required, such as spinning the fine counts, these may be replaced by pads made from cork, compressed fibre, or other material. If the friction of the material used for the pad is high then this automatically leads to a high frictional torque. A low friction material leads to low frictional torque and its corollary, low spinning tension. The friction of a felt pad can become greater if it becomes contaminated with grease or dirt. The carrier sleeve/dead spindle bearing must be greased and for this reason a small grease-cup is formed at the upper end of the dead spindle. Each fresh charge of grease melts as the carrier rotates around the spindle and generates heat, and the grease runs down the spindle. After a time (or with excessive greasing) the grease finds its way on to the drag-pads and in so doing increases the friction between them and the bearing plate. For this reason, greasing should be carried out carefully, and periodically the drag-pads need to be cleaned with solvent.

With well-maintained felt pads the coefficient of friction can be as low as 0·6, leading to an average tension in the yarn between the flyer eye and the bobbin of about $1\frac{1}{4}$ lb. If, however, the pads become contaminated with grease and dirt the coefficient of friction may rise as high as 0·9, under which circumstances the tension will be around $1\frac{3}{4}$ lb.

So far, spinning tension has been considered in general terms but it is now necessary to discuss it in greater detail as it is one of the prime factors in determining how well the frame will perform and what its production capabilities will be. Spinning tension and yarn breaks are closely related and since the repairing of yarn breaks is the chief duty of the spinner, their number will determine, to a very large extent, the workload of the spinner and the labour requirements at the spinning stage. Spinning tension arises from the method adopted for winding-on in jute frames. It was shown earlier in this chapter how the frictional torque steadily grew as the bobbin filled up but, because of the faster radial increase of the bobbin, the spinning tension fell during the course of each doff. Spinning tension, however, is not of the same magnitude in all parts of the yarn from the bobbin surface up to the drafting nip.

Two levels of tension are found, on-winding tension and transmitted tension. On-winding tension is the tension developed in that part of the yarn between the flyer eye and the surface of the bobbin. Transmitted tension is the tension in that part of the yarn above the wharf-cap. The transmitted tension is *always* lower than the on-winding tension. The way that the yarn tension varies from the

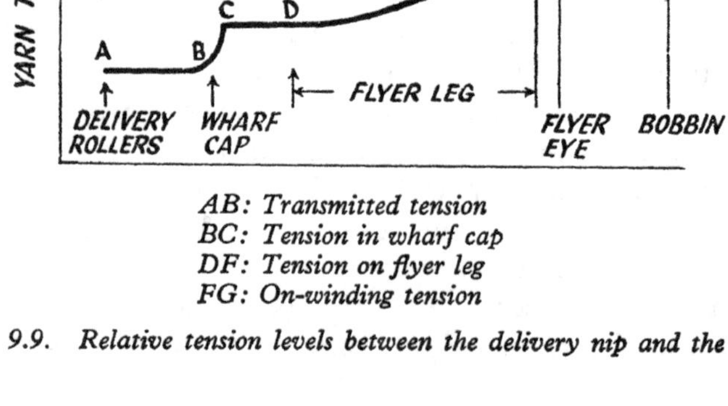

AB: Transmitted tension
BC: Tension in wharf cap
DF: Tension on flyer leg
FG: On-winding tension

Figure 9.9. Relative tension levels between the delivery nip and the bobbin

drafting nip down to the bobbin is shown in Figure 9.9. The gradual reduction in tension as one progresses back up the yarn from the bobbin is due to the well known capstan effect. A ship can be moored to a jetty merely by wrapping its rope a number of times around a capstan or bollard, no knot being required to keep the boat from drifting away as the friction between the rope and the bollard is sufficient to keep the vessel secure. On the spinning frame, the on-winding tension can be regarded as equivalent to the tension between the vessel and the bollard and the transmitted tension equivalent to the free end of rope lying on the quay-side. The level of the on-winding tension is determined solely by the rotational torque required to overcome the frictional torque of the drag-pads; the transmitted tension depends upon on-winding tension *and* the friction between the yarn and the flyer leg and the length of yarn in contact with it. The length of yarn in contact with the leg depends upon the number of times it is wrapped around it in its downward passage from the top of the leg to the flyer eye and the size of the other small angles where it bears against the ceramic disk at the foot of the hole through the wharf and against the flyer eye. The relationship between the on-winding tension and the transmitted tension is given by

$$T_t = T_o \exp(\mu\theta)$$

where T_t is the transmitted tension, T_o the on-winding tension, μ the coefficient of friction of jute yarn on steel, and θ the total angle of wrap on the flyer leg and any other bearing surfaces.

Since T_o is fixed by the frictional torque at the drag-pads and μ is constant, or nearly so, for all normal jute yarns it follows that the only way in which T_t, the transmitted tension, can be altered is by changing the angle of wrap on the flyer leg. Plate V shows three methods by which the yarn can be led from the top of the flyer to the eye. In practice, only the straight-through and the once-round thread-up is used, for if the yarn is wrapped twice round the leg the transmitted tension is so low that ballooning usually occurs. On a 4¼ in. pitch frame, the average on-winding tension is commonly of the order 1¼ lb and the transmitted tensions for straight thread-up, once round, and twice round the leg are about 0·5, 0·25, and 0·15 lb. It will be appreciated that if the straight thread-up is used the yarn will be subjected to a higher tension throughout its passage from the drafting nip to the flyer eye; this may give rise to a slight increase in spinning breaks. It will usually be found that when the heavy counts are being spun the spinners use the straight thread-up so that the transmitted tension will be sufficiently high to prevent ballooning.

On the 4¼ in. frame, a popular choice for hessian yarns, the on-winding tension at the start of the spin is of the order 1·75 lb and drops to about 1·0 lb at the end, although wide variations in these values are found. The transmitted tension usually begins around 0·3–0·4 lb then falls steadily to 0·15–0·20 lb. In spite of all that has been said so far about the influence of spinning tension on the number of end-breaks that occur, it might be thought that these levels of tension are far below the average strengths of jute yarns. It is not, however, the *average* strength that is important in this respect but the *minimum* strength. For 8 lb/sp yarn this may be around 2 or 3 lb—still apparently well above the spinning tension level and the reason for the correlation between breaks and tension is not brought to light until one examines the tension by means of high-sensitivity instruments which have a rapid response to sudden tension pulses of extremely short duration. Then it is found that while the general level of tension, on-winding and transmitted, is set basically by the frictional torque at the carrier base and the angle of wrap round the flyer leg, there are irregular, sudden tension pulses of extremely short duration which are many times as great as the average level. These are the cause of yarn breaks in spinning.

Basically, they are all connected with bobbin rotation. The bobbin sits upon a carrier which, in turn, is set with a loose fit on the dead spindle. In addition, the surface of the bobbin is not smooth and the effective bobbin radius is changing in an irregular manner as the yarn builds momentarily on top of a yarn in the previous layer then at the next instant falls into a groove between two coils of yarn; with this continual change in radius it follows that the torque supplied to the bobbin is altering from moment to moment. Under these circumstances it would be rather surprising if the bobbin and the carrier rotated smoothly round the spindle. In fact, the motion of the bobbin is subject to a series of sudden accelerations and decelerations which cause the tension in the yarn to be jerky and irregular. Any defect in the spindle/carrier/bobbin assembly is liable to accentuate these irregularities in rotation and consequently leads to higher and more frequent tension pulses in the yarn, with the certainty that more ends will break as a result.

One defect that is inherent in the design of the spindle and the way the spinning frame operates is that the spindle can only be supported at one end. When the bobbin is at the bottom of its traverse the yarn is winding on at the top of the bobbin and is, as it were, 'pulling' the bobbin sideways at the top and exerting a force which tends to 'bend' the spindle. The spindle, of course, does not deflect but it does vibrate more violently than when the yarn is winding on at the foot of the bobbin and the leverage from the winding point to the attachment of the spindle is short. Plate VI shows a high-speed record of the spindle vibration with a short piece of the transmitted tension record attached. The trace refers to one complete builder cycle and the width of the band gives a measure of the degree of spindle vibration; it will be seen that more vibration occurred in the middle of the trace when the spindle was withdrawn from the flyers. The tension trace shows how the tension in the yarn rose as the spindle vibrated more violently.

Another cause of tension pulses is the jerk the flyers receive each time the joint in the driving tape comes on to the wharf; no matter how small and neat the join, this always happens. This jerk imparts a sudden pulse to the tension in the yarn as Plate VII shows; each vertical line arising from the sudden change in the flyer velocity as the joint of the tape comes round. These peaks are of the order of 1·5 lb, roughly 5 times as great as the average transmitted tension.

If the spindle is not exactly central with reference to the flyers, or if it is not vertical and straight, then the rotation will be more irregular than necessary and consequently not only will there be more tension pulses than normal but they will also be more vigorous and yarn breaks will be increased. Yet another cause of irregular spinning tensions is bad bobbin building; if the builder is not aligned properly with reference to the flyers the bobbins will be under-built at one end and over-built at the other. This is undesirable since the narrower bobbin radius will bring about a rise in the average tension and also cause more tension pulses, for the bobbin surface at these points is always more irregular than normal. Needless to say, if the drag-pads are greasy or contaminated in any way they will have a greater tendency towards stick-slip rotation with the consequent irregular tension it sets up.

Figure 9.10 shows a selection of spinning tension traces taken of the transmitted tension on a $4\frac{1}{4}$ in. pitch frame; it should be noted that these traces, because of the response of the instrument, only show the general levels of tension and do not reveal the short-duration pulses that have been discussed above. However, if the general level of tension

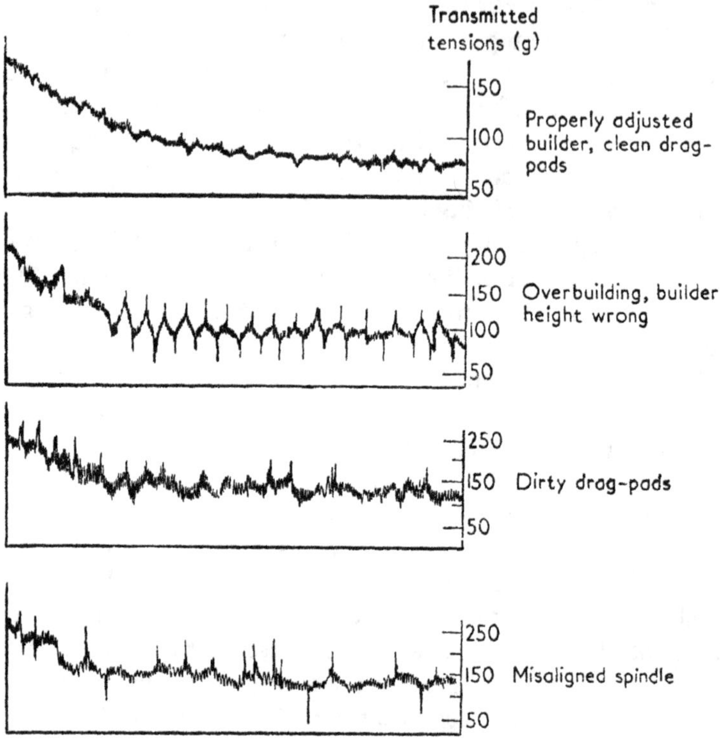

Figure 9.10. Spinning tension records. 4¼ in. frame, 276 tex yarn, 4,000 r.p.m.

is high or irregular, then the rapid tension pulses will be even higher and more frequent.

Spinning tension and frame speed are two closely related features. It has been known from the earliest days of spinning that if the frame speed is increased more yarn breaks occur; the form of the variation of breaks with speed is shown in Figure 9.11. When tests were made of the average tension in the yarn at different frame speeds, however, no change in the general tension level could be seen and it was only when the tension pulses were examined that the reason behind the increased number of yarn breaks was found. Table 9.2 shows the results of one such test. Clearly, at the higher frame speeds many more tension pulses occur.

At flyer speeds of 4,250 r.p.m. the tension peaks of 300 g are about 4 times as frequent as at 2,160 r.p.m. and peaks as high as 650 g are found. When one remembers that these records were taken above the

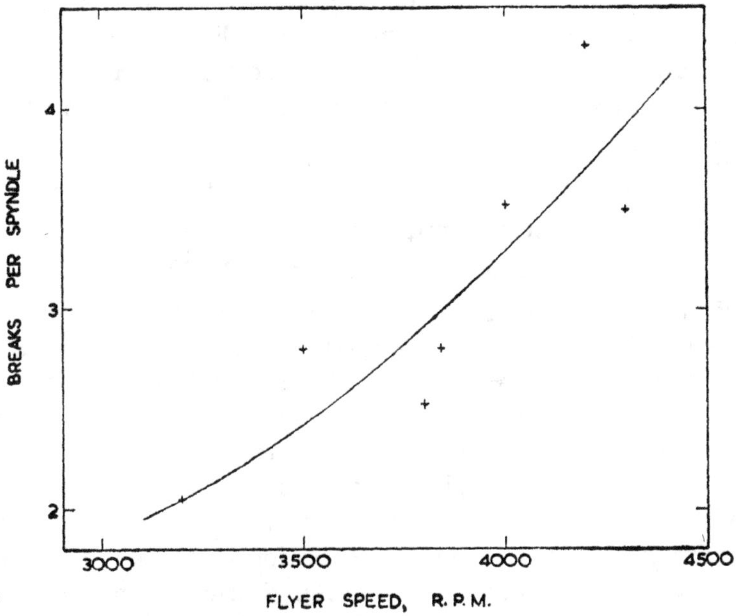

Figure 9.11. Effect of flyer speed on spinning breaks, 276 tex hessian warp

wharf cap it will be appreciated that the *on-winding* tension pulses are of quite a high magnitude (occasional ones being as high as 5–6 lb). Tension pulses of this order are much greater than the weakest parts of the yarn can stand and whenever a pulse and a weak spot coincide the yarn will break.

Under European conditions one spinner can attend to about 200 spindles when hessian yarns are being produced at between 3,700 and 4,000 r.p.m., but greater or lesser spindle allocations are found

TABLE 9.2. FREQUENCY OF TENSION PULSES ON A $4\frac{1}{2}$ IN. FRAME

Transmitted tension (g)	*Pulses per minute*	
	4,250 r.p.m.	*2,160 r.p.m.*
300	3,960	860
400	256	10
500	12	0
600	1·8	0
650	0·6	0

depending on the grade of fibre being worked. The spinning breakage rate varies widely from mill to mill since quality, frame maintenance, and tension levels differ, but a general indication of the breaks per 100 spindles per hour when spinning 260–300 tex yarns at about 4,000 r.p.m. is

Hessian warp	20–45
Hessian weft	30–55
Carpet or linoleum yarns	20–30

Virtually all spinning breaks are caused by the spinning tension exceeding the yarn strength, the exceptions being caused by the yarn striking an adjacent flyer when it balloons off a flyer leg or when one end breaks and becomes tangled with its neighbour, causing it to fall. The number of yarn breaks during spinning has been shown to be closely associated with spinning tension and yarn quality. If the number of end-breaks is large then a greater work-load is thrown on the spinner and it will become impossible for the operative to repair all the breaks as they occur. Consequently there will be some ends permanently down on the frame. The number of spindles that are idle as a result of end-breaks depends not only upon the frequency of the breaks but upon the skill and diligence of the spinner. It is common experience that a skilful spinner can cope with a reasonable number of end-breaks without having too many ends idle, 2 or 3 per cent of the total spindles being a typical figure for average break rates, but a poorer spinner may have 7 or 8 per cent of the ends idle under similar conditions. Thus the skill of the spinner can have a great influence upon the output from a frame.

The rate at which the ends break is not constant throughout the spin, being greater at the start when the spinning tension is high, but, in addition, there is another effect seen at the very start of the bobbin, as the following figures show.

1st builder cycle (approx. $1\frac{1}{2}$ min)	20 per cent of all breaks
Remainder of first 5 min	10 per cent of all breaks
Remainder of bobbin	70 per cent of all breaks

Extensive tests on 8 lb/sp hessian warp and weft yarn have shown that 20 per cent of all the yarn end-breaks occur during the first cycle of the builder when the yarn is winding on to the bare core of the bobbin or one layer of jute. At this time there is no resilience to the bobbin surface and consequently any sudden acceleration or deceleration of the bobbin results in a very high tension peak; later in the spin there is a pad of jute to help to absorb some of the impulsive loads that are thown on to the yarn. Again some 10 per cent of the breaks occur during the rest of the first 5 min of the spin, a time when the spinning tension is high. These high break-rates throw a heavy load on the spinner at the start of each doff and it is always found that the number of idle spindles is greater at the start of the bobbin than during the remainder. Another source of high starting breaks is the careless use of the starting handle of the frame. If the operative starts the frame with a sudden jerk then a large number of ends will certainly fall and, for this reason, the frame should always be started gently.

PRODUCTION ASPECTS OF SPINNING

The output from a spinning frame depends upon the yarn count, twist, and flyer speed.

Jute spinning frames are made in several sizes, each to suit a certain range of counts. Table 9.3 shows the operating details of the various sizes of frame.

TABLE 9.3. SPINNING FRAME DATA

Frame pitch (in.)	*Bobbin dia. (in.)*	*Bobbin length (in.)*	*Count range (lb/sp)*	*Flyer speeds (r.p.m.)*	*Wt of yarn on bobbin (lb)*	*Number of spindles*
$3\frac{1}{2}$	$2\frac{3}{16}$	$4\frac{3}{4}$	$3\frac{1}{2}$–6	4000–4200	0·27	110
$4\frac{1}{4}$	$2\frac{11}{16}$	$5\frac{1}{2}$	6–10	3700–4200	0·50	100
$4\frac{1}{2}$	$2\frac{7}{8}$	$6\frac{1}{2}$	10–18	3000–3600	0·65	90
$5\frac{1}{2}$	$3\frac{1}{2}$	$7\frac{1}{4}$	16–30	1900–2500	1·15	80
6	4	8	20–48	1500–2200	1·75	70

These figures are a general guide and it is quite often found that the count range will be extended somewhat to suit the sales requirements in any one mill.

If P is the frame production, v the delivery speed, n the flyer speed, t the yarn twist, k the twist factor, η the frame efficiency, and c the yarn count,

then,

$$P = cv\eta$$

but,

$$v = \frac{n}{t}$$

and,

$$t = \frac{k}{\sqrt{c}}$$

hence,

$$P = \frac{nc^{3/2}\eta}{k}$$

Thus high production will be achieved with high flyer speeds, high counts, high efficiencies, and low twist factors. Of these variables, k and c are fixed by sales requirements and so from the practical point of view it is only necessary to examine n and η.

FLYER SPEED

The speed at which the frame can be run depends firstly on the ability of the yarn to spin successfully without breaking too often, and secondly on the mechanical capabilities of the machine itself. It has already been indicated that as the speed of the frame is increased then more tension pulses arise and consequently more end-breaks occur. The better qualities of yarn are better able to withstand the stresses of high speed and it is always found that the lower the yarn quality the more end-breaks take place.

With the present design of flyer spinning frame there are limitations put upon the upper limits of speed by the performance of the flyers themselves. These are largely connected with the throw-out of the legs due to centrifugal force. Throw-out depends upon several physical factors. For example,

$$\begin{aligned} \text{throw-out} &\propto \text{cross-sectional shear strength of the leg} \\ &\propto \text{leg separation} \\ &\propto \text{speed}^2 \\ &\propto \text{leg length}^4 \end{aligned}$$

It will be seen that the two most important factors are flyer speed and leg length, for example an increase in speed from 3,000 to 4,200 r.p.m. gives twice as much throw-out and changing the leg length from 6 to 7 in. would raise the throw-out by the same amount.

The amount of throw-out assumes very great importance in determining the size of the bobbin that can be used since all flyer legs are inclined inwards so that when the frame is running and throw-out takes place the legs will assume a vertical position. The inward inclination limits the size of the bobbin diameter that can be used—the bobbin diameter must always be slightly less than the leg separation when the flyer is at rest, otherwise the flyer would jam on it each time the frame was stopped. Therefore, if (1) high speeds or (2) long bobbins are to be used then this increases the throw-out which, in turn, limits the maximum bobbin diameter that may be employed. It will be apparent, therefore, that the yarn carrying capacity, package size, and flyer design are interdependent, and similarly delivery speed, bobbin rotation, and the ability of the yarn to withstand spinning tension are interdependent. It would seem that any twisting device that could overcome these restrictions, or at least some of them, might offer a possibility of increased spinning production. Several flyer designs have been patented, the main objective being to limit throw-out and give either higher flyer speeds or larger spinning packages or both. Flyers resembling open-sided cylinders with a complete ring at the foot and flyers made in two wing-like halves have been used with some success, though the solid ring at the foot does interfere slightly with the normal processes of repairing end-breaks and cleaning. As a development from the wing-like flyers, a counterbalanced single-wing flyer has been produced by the Fairbairn Lawson Textile Machinery Co. Ltd called the Falaflyer. In comparative tests between the Falaflyer and more conventional types it has been found that fewer tension peaks occur. For example, when running at 4,000 r.p.m. a conventional flyer gave 785 pulses greater than 320 g above the cap during the first 5 min of the spin. Under the same conditions the Falaflyer gave only 125. The Falaflyer is combined with the double apron draft control unit, Figure 9.1 (d), and a motor-driven traverse on the Falaspin frame which, it is claimed, gives a stronger, more regular yarn with fewer ends down. The shape of the Falaflyer allows a larger bobbin to be used with an increase of 40 per cent in yarn-carrying capacity compared with the conventional size for a $4\frac{1}{4}$ in. frame.

Another interesting method of simultaneously twisting and winding-up the yarn is found in centrifugal spinning or, as it is more commonly known, 'pot spinning'. Centrifugal spinning has been known from the beginning of the century when Topham produced his first 'pot' for spinning viscose rayon. With the wet-spun viscose there was little difficulty in producing a stable package which could be handled in subsequent processes, but it was not until 1948 that the first commercially available machine was released for use with dry yarn. This machine was the Prince-Smith Centrifugal (P.S.C.) machine for worsted yarn. The general arrangements of the centrifugal spinning system are shown in Figure 9.12. The yarn descends into a rapidly

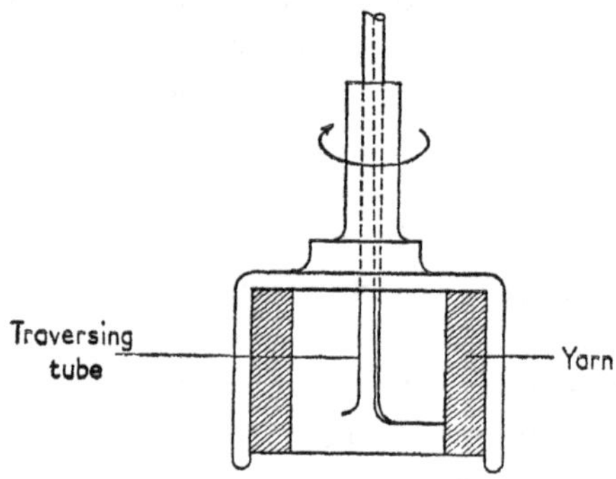

Figure 9.12. Diagrammatic representation of centrifugal spinning

rotating pot and when it comes against the inner surface of the pot it simultaneously becomes twisted and wound-on by centrifugal force. The yarn continues to wind the package from the outside to the inside, then when the packing is full a spring cage or some other device rises into the pot and withdraws the yarn. This type of spinning has great speed potential and while the practical application of the system has not been fully exploited commercially for jute it will be interesting to watch the future developments along these lines.

SPINNING EFFICIENCY

From the viewpoint of production there are several sources of lost time at the spinning frame; doffing, end-breaks, maintenance, lack of orders, etc. This allows one to formulate different levels of efficiency.

(1) Spindle efficiency, the number of spindles that are actually

producing on the frame. Those which are idle because the ends have broken and the spinner has not yet repaired them are deducted from the total number on the frame; usually it is expressed as a percentage, e.g. a spindle efficiency of 96 per cent means that out of every 100 spindles 96 are actually producing and 4 are idle.

(2) Frame efficiency, the actual running time of the frame expressed as a percentage of the total possible running time. Note that this takes account of time lost through doffing *and* spindle efficiency, e.g. a frame with 100 spindles is producing 8 lb/sp yarn at 26 yd/min, the bobbin holds 0·5 lb of yarn and the frame is stopped for doffing for 70 sec, on average there are 2·8 spindles idle because of end-breaks; the frame efficiency is then

0·5 lb of 8 lb/sp yarn	900 yd
Time to spin 900 yd	34·6 min
Doffing time	1·16 min
Total time to fill and doff the bobbins	35·76 min
Number of spindles working effectively	= 100–2·8
	= 97·2

$$\text{Frame efficiency} = \frac{34{\cdot}6 \times 97{\cdot}2}{35{\cdot}76 \times 100} \times 100 \text{ per cent}$$
$$= 94 \text{ per cent}$$

(3) Flat efficiency, this takes account of the frame efficiency *and* lost time through maintenance, spinners' absence, lack of orders, etc. This value for efficiency gives, as it were, a measure of how effectively management is using the productive capacity of the spinning department.

For example, a spinning department of 30 frames has, on average, frame efficiency of 90 per cent and during one particular 40 hr week, 6 hr are spent on maintenance, 1 frame stood for 4 hr because the spinner was ill, and 2 frames were off for 1¼ days through lack of orders. What was the flat efficiency?

Total possible time = 30 × 40 frame hours

Lost time:

Maintenance	6 hr
Spinner's absence	4 hr
Lack of orders (2 × 10)	20 hr
	—
	30 hr

Flat efficiency

$$\frac{(1200-30)\times 90}{1200\times 100}\times 100 = 87{\cdot}6 \text{ per cent}$$

STOP MOTIONS AND PIECING

There is only one stop motion on a jute spinning frame and this is for detecting a broken end of yarn. A light porcelain finger rests against each yarn between the drafting rollers and the wharf cap. When the yarn is running normally the finger is held back but when the yarn has broken it allows the finger to swing forward and, through a system of levers and trip-rods, the front retaining roller springs away from its fellow and the rove or sliver stops passing between them. This front roller is held against the driven retaining roller by means of a heavy counterweight which drops forward when the stop motion operates. This weight not only provides the energy required to separate the two retaining rollers but acts as an indicator and shows the spinner that an end has broken.

To repair the break, the spinner grasps the wharf cap firmly and stops it (this can be done quite safely because the driving-tape to the wharf slips on the polished surface). A special wire hook is then passed down through the hole in the wharf cap and the broken end of yarn on the bobbin drawn up through the cap. The spinner grasps this end in the right hand and quickly places it into the nip of the drafting rollers. Simultaneously, the spinner lets the wharf cap go and the flyer immediately rotates. With the left hand the spinner swiftly pushes the counterweight upwards, making the retaining rollers engage once more and bring down the rove or sliver. As soon as the rove or sliver reaches the drafting zone the broken end of yarn is released and passes through the drafting nip along with the fresh supply of sliver or rove, becoming twisted in with it as soon as it emerges from the nip and comes under the influence of the flyer. In this way a broken end can be joined on to a new piece without a knot—the whole operation being known as piecing or splicing. Inevitably, a splice is about twice the normal thickness of the yarn because the broken end must be twisted in with a new piece if the two are to hold together. This is a defect in the yarn, but one which is unavoidable—all that can be done in this respect is to try to keep the splices as small and neat as possible. As may be imagined, the operation of piecing requires considerable skill and experience before a neat effective splice can be made each time. A good spinner can carry out a splice in ten seconds or even less but a poor spinner may take longer and find that the end breaks when the splice passes down the flyer leg.

A yarn defect, known as 'spinner's double' may arise if the stop motion is not functioning properly. If the supply of material is not cut off quickly enough when an end breaks then a ribbon of drafted, but untwisted, jute is liable to drift across to its neighbour and become twisted with it, producing a double-count portion of yarn which may extend for some distance.

CHAPTER TEN

The System

IN THIS chapter, the factors which must be considered when assembling the various stages of the process into a complete plant will be dealt with. If an entirely new range of machinery is bought from one maker then he will ensure that the machines work smoothly as a whole and that each stage is operating under the best conditions for the grade of raw material envisaged. It is more likely, however, that one stage in an existing plant is to be replaced, or production requirements have changed in some way since the plant was installed. Under these conditions it is vital to have a knowledge of the fundamental factors governing the manner in which the separate stages may be integrated to form an efficient unit. Only then can the full potentialities of the machinery be realized.

A certain amount of the material in this chapter has been discussed earlier, but it is felt that for the sake of completeness, a reappraisal at this stage of some of the salient facts would not come amiss.

PRODUCTION ASPECTS OF THE SYSTEM

The term 'system' in the production sense refers to the integrated sets of machines fed from one breaker card, but in this chapter the term has been carried one stage back to the spreader or softener so that the inter-relationships of the complete set of machines may be studied.

If the system is to work satisfactorily it must conform to several conditions

(1) It must be able to produce the correct count of yarn.
(2) At no stage must the sliver be excessively heavy or light for the range of machinery in use.
(3) It must be capable of high production rates and operate with as low a labour force as possible.
(4) Each stage in the system must be able to produce enough material to satisfy the succeeding one; similarly, it must be able to consume all the material put out by the preceding one, i.e. the system must be 'balanced'.

By successive drafting the sliver count must be reduced to a level suitable for the spinning frames to work on. Since the spinning frames can only operate within a certain draft range the count of the finisher drawing sliver must be such that all the desired range of counts can be spun from it. The figures in Table 10.1 represent the normal range of sliver counts for hessian yarns.

TABLE 10.1

Sliver	*ktex*	*lb/100 yd*
Spreader	220–320	45–65
Breaker card	85–105	17–21
Finisher card	65–90	13–18
1st drawing	40–80	8–16
2nd drawing	23–28	4·5–5·5
Finisher drawing	4·5–6	0·9–1·2

For heavy yarns where there are only two drawings in the system the card slivers may be some 10 per cent heavier. The sliver from the first drawing is usually in the range 35–45 ktex and that from the finisher drawing 8–10 ktex.

These sliver weights are obtained from a variety of draft and doublings combinations but the figures in Table 10.2 represent typical arrangements for the production of hessian yarns.

TABLE 10.2

Machine	*Draft*	*Doublings*
Spreader	8–12	2 leaders (optional)
Breaker card	15–20	6–8
Finisher card	10–14	10–12
1st drawing	3½–5	2 or 4
2nd drawing	5–7	2 or 3
Finisher drawing	8–10	1 or 2

To achieve the maximum output from a given set of machines it is necessary to work with as high delivery speeds as possible. There is however a limit to the speed at which machines can be driven; above this level the machine will not function in the proper manner and the product is unacceptable or stoppages due to mechanical troubles become too frequent or both these defects arise simultaneously.

In addition, the count of the material must not overload the machine for which it was intended otherwise the necessary operations of splitting, opening, cleaning, parallelizing the fibres, and producing a regular sliver and yarn will not be carried out effectively, and the quality will suffer. On the other hand, if the sliver is too light for the range of machinery it is likely that short fibre control during drafting will be ineffective as the pins, rollers, aprons, etc., will be underloaded. There is, moreover, another important feature which determines the lower level of sliver count and that is the ability of the material to unwind from a roll or withdraw from a can. Nowadays, all spreader sliver and most card sliver is handled in roll-form and it has been found that difficulty is experienced in unwinding if the count of the sliver is low. This applies particularly at the spreader where the count of the lightest parts of sliver may only be about one-third of that of the average. Similarly sliver in card rolls must be strong enough to withstand the slight tension that must be applied between the take-off gear and the feed rollers to stop the slivers sagging. At the later stages there must be sufficient sliver bulk to give strength and cohesion for the sliver to be withdrawn from its cans without parting. It will be seen, therefore, that the lower limit of sliver count is intimately bound up with packaging, and since large packages are an essential economic

Speeds of Jute Machinery (ft/min)

Spreader delivery	
Cards:	180–225
Breaker, cylinder	2,400–2,700
Breaker, delivery	150–200
Finisher, cylinder	2,400–2,800
Finisher, delivery	150–210
Drawing frames, faller drops:	
1st, push-bar	700–850
2nd, spiral (double)	250–400
Finisher, spiral (triple)	350–650
Spinning, flyer r.p.m.:	
7–12 lb/sp	3,750–4,250
13–24 lb/sp	2,800–3,600

feature of jute spinning (especially in high labour cost countries) it is one which assumes considerable importance.

The speed and loading at which the machines may be run are dependent upon the design of the machine, the density of pinning, etc., but the following figures represent average values for hessian yarns and the like.

Average Machine Loadings, lb/100 yd

Spreader	500–700 on the gill-pins
Breaker card	350–400 on the feed sheet
Finisher card	170–240 on the feed sheet
1st drawing	14–18 on the gill-pins
2nd drawing	8–10 on the gill-pins
Finisher drawing	8–10 on the gill-pins

It will be appreciated that these conditions impose certain restrictions on the method of operating an integrated set of machines but, besides this, there is another important factor which must be carefully considered—the flow of material from one stage to the next.

Before this is discussed, however, it is necessary to consider machine efficiency. 'Efficiency' is a term which can cause confusion since it can mean different things to different people. Efficiency is commonly related to performance in the following way

$$\text{efficiency} = \frac{\text{machine running time}}{\text{machine running time} + \text{stopped time}} \times 100 \text{ per cent}$$

The confusion arises from the definition of stopped time. Stopped time may be taken as that time during which the machine is stopped for doffing but it may also refer to all the lost time during working hours and include maintenance time and time lost when the machine is waiting for material to work on. Taken over short periods the doffing time may be all that is allowed for and a high figure for efficiency will be obtained, but this is not a measure of the real operating performance of the machine when it is integrated in a system. Then, its efficiency is governed by doffing *and* the availability of raw material. For instance if machine A is producing material at a rate of 20 yd/hr for machine B and the latter can consume it at 30 yd/hr, then B can only work for

$$\frac{20}{30} \times 100 \text{ per cent}$$

of the time, since A has only 20 yd of material available in an hour. The efficiency at which B will operate is therefore closely bound up with A's performance. In this way in the jute processing system the efficiency of each machine is governed by the performance of the machine immediately before and after it in the processing sequence.

Again, the pursuit of high *efficiency* alone is not always justifiable since under certain circumstances a higher *output* can be realized by running the machine at a faster speed, even if this is accompanied by a lower efficiency, and *vice versa* (Table 10.3).

TABLE 10.3

Delivery speed	*Efficiency*	*Output*
10 *yd/min*	90 per cent	9 *yd/min*
12	88	10·6
14	85	11·9
16	80	12·8
18	70	12·6
20	50	10·0

The manner in which the separate machines in the jute processing system combine to form an integrated whole is of interest and has important practical implications since an examination of the inter-relationships between the various stages shows that a departure from the balance planned by the machinery-maker may lead to uneconomic operation. The prime factor to be considered in any system is the flow of material from one stage to the next, for it is this which governs the efficiency, the speed, and the number of machines required at each of the several stages.

Consider the finisher cards and the 1st drawings of a certain system; if the cards can produce more sliver than the 1st drawing frames can consume then, inevitably, there will be a build-up of material. To prevent this growing continually, the cards must be stopped for a time to let the drawing frames work away the accumulation of sliver. On the other hand if the drawing frames consume sliver faster than the cards can produce it then *they* must be stopped to allow the cards to work up a stock of sliver to supply the drawing frames. Ideally, the cards' delivery should be exactly equal to the drawing frames' feed. This situation can rarely be achieved because of the different speed capabilities, doffing requirements, and maintenance times between two successive stages. Under the practical conditions the stages automatically adjust their efficiencies so that the total net output at one stage equals the total net input at the next. This must happen in all systems otherwise there would be an accumulation of material somewhere in the process.

In a system there may be three finisher cards, each with a delivery speed of 200 ft/min and 20 drawing frame feed slivers running at 25 ft/min

$$
\begin{aligned}
\text{Total card output} &= 3\times 200 \\
&= 600 \text{ ft/min} \\
\text{Total drawing frame input} &= 20\times 25 \\
&= 500 \text{ ft/min}
\end{aligned}
$$

If all the machines worked continuously then there would be a card sliver gain of 100 ft/min over the drawing feed. But if the cards worked for only 5/6 of their time they would produce the necessary 500 ft/min of sliver for the drawing frames. In practice, however, the drawing frames cannot run non-stop and their efficiency will only be perhaps 80 per cent, i.e. their feed requirements are no longer 500 ft/min but 400 ft/min (80 per cent of 500). The finisher cards need only work long enough to provide this quantity of sliver and if they operate for 66·7 per cent of the time they will do this; in other words, their efficiency need only be 66·7 per cent to satisfy even flow conditions. If, for any reason, the drawing frame efficiency alters, then the card efficiency must alter too in order that no build-up or deficiency of material will arise.

At each transfer in the process from one set of machines to another the output and input are linked in this manner and the machine efficiencies are mutually interdependent. The general equations for transfer are

$$p_o = v_o n_o \eta_o$$

$$p_i = v_i n_i \eta_i$$

where p is the production, v the machine speed, n the number of slivers at each stage, and η the machine efficiency, and the subscripts o and i refer, respectively, to the output at one stage and the input at the next stage. For even flow $p_o = p_i$ and the equations become

$$v_o n_o \eta_o = v_i n_i \eta_i$$

By means of these equations it is possible to calculate the efficiency at each stage in the process if the delivery speeds, drafts, and doublings are known. Once the operating conditions of draft, speed, etc., have been chosen then the machine efficiencies are predictable and will only be departed from over short periods of time as the flow of material oscillates between overproduction and underproduction.

The operating data for a system producing 600 kg/hr of 276 tex yarn are given in Table 10.4.

TABLE 10.4

	Number of deliveries	*Draft*	*Doublings*	*Delivery speed (m/min)*
Spreader	1	10·0	—	48
Breaker cards	2	17·4	6	61
Finisher cards	3	12·0	10	50
1st drawings	4	4·0	4	46
2nd drawings	24	6·5	2	24
Finisher drawings	60	9·0	2	47
Spinning frames	1600	18·0	—	25

The first step in setting up a list of efficiencies is to calculate the 100 per cent delivery and feed at each stage. One stage is then selected as the key stage from which the other efficiencies will stem. The key stage may be at the start of the process or at the end or even in the middle, the choice being made from some prior experience. In general, spinning frames operate with an efficiency of 85–90 per cent, drawings at 70–80 per cent, cards at 80–90 per cent, and spreaders at 70–80 per cent. The arithmetic of the calculation will only be shown for one transfer, all others being carried out in the same manner.

Consider the transfer from the spreader to the breaker cards, then from the flow equations

$$v_o n_o \eta_o = v_i n_i \eta_i$$

$$48 \times 1 \times \eta_o = \frac{61}{17 \cdot 4} \times 2 \times 6 \times \eta_i$$

$$\eta_o = \frac{61 \times 12 \times \eta_i}{48 \times 17 \cdot 4}$$

$$\eta_o = 0 \cdot 88 \eta_i$$

If now several hypothetical spreader efficiencies are assumed then the corresponding breaker card efficiency can be calculated from this relationship, e.g.

Spreader efficiency	*Breaker card efficiency*
70	80
75	85
80	91

The complete analysis of the *system* efficiency for several different *spinning* efficiencies is shown in Table 10.5.

TABLE 10.5

Spinning frames	88	90	92	94
Finisher drawings	69	71	73	76
2nd drawings	74	76	77	79
1st drawings	72	74	75	77
Finisher cards	88	90	92	94
Breaker cards	91	93	95	98
Spreader	80	81	84	86

It will be noted that when the spinning efficiency rises above 92 per cent the early machines must operate at very high efficiencies and it is doubtful if such performance could be maintained for long.

The importance of knowing the system flow characteristics cannot be overestimated for without such knowledge the full potentialities of the system cannot be appreciated. It is a simple matter to measure all the surface speeds in the mill and count the number of slivers at the feed and delivery of each stage; this is all the information necessary for such an analysis. By means of such calculations it is possible to find the effect of working overtime in one department or the effect of changing the speed or draft of any machine. Finally, one practical point about such calculations—it is always preferable to work in terms of the length delivered and consumed at each stage. It is possible to use the weight of sliver produced but, for accuracy, this requires that allowances be made for waste and moisture losses and if the length method is used then such complications are avoided.

An illustration of how a system may be analysed will be given to show the general method adopted. It should be noted that in an analysis of this type there is not one solution but a number, each of which is equally suitable provided they fulfil the conditions of speed, loading, and attainable efficiency at each stage. The analysis will be carried out on a hypothetical system which has become unbalanced, the object being to improve the system as much as possible.

TABLE 10.6

Stage	*Sliver Count (lb/100 yd)*	*Delivery speed (ft/min)*	*Draft*	*Doublings*	*Number of deliveries*
Spreader	65·0	171	10·0	—	1
Breaker cards	23·0	186	19·8	7	2
Finisher cards	20·9	156	11·0	10	3
1st drawings (push-bar)	12·0	180	3·5	2	6
2nd drawings (spiral)	5·7	99	6·3	3	20
Finisher drawings (spiral)	1·14	114	10·0	2	72
Spinning	0·056	78	20·4	—	1,800

First, the existing system is examined and the stage efficiencies calculated by means of the basic flow equations. Next, the sliver counts are studied and the loadings on the feed sheets and gill-pins at each stage found. Then the machine speeds are examined, with particular reference to the faller drops per minute at the drawing frames. In the system above, a general reduction in the sliver count early in the process would be likely to improve the yarn quality since the machinery is either working at its maximum loading or just over it.

TABLE 10.7

Stage	*Efficiency (per cent)*	*Faller drops*
Spreader	53	—
Breakers	69	—
Finishers	71	—
1st drawings	55	1,330
2nd drawings	63	400
Finisher drawings	70	375
Spinning	90	—

The 1st drawing frames are operating with too high a rate of faller drops, while the rate of the finisher drawing could be substantially increased. If the speed of the first drawing is reduced this will bring the faller drops to a reasonable level; this should be possible since there is scope for running the machines with a much higher efficiency than 55 per cent. Similarly, if the finisher drawing frames are speeded up it should be possible to reduce the number of deliveries, thus effecting savings in running costs.

The arithmetic of each step will not be shown since it only involves the basic flow equations and the calculation of the number of faller drops per minute dealt with in an earlier chapter. It will usually be found that several sets of calculations must be made, the final solution being obtained by selecting suitable data from each set of results. One method of revising the system is shown in Table 10.8.

TABLE 10.8

	Speed (ft/min)		*Sliver count (lb/100 yd)*		*Number of deliveries*		*Efficiency (per cent)*	
	E	*R*	*E*	*R*	*E*	*R*	*E*	*R*
Spreader	171	171	65·0	51·0	1	1	53	80
Finisher cards	186	186	23·0	21·0	2	2	69	91
Breaker cards	156	150	20·9	17·5	3	3	71	90
1st draw.	180	120	12·0	10·0	6	6	55	86
2nd draw.	99	99	5·7	5·5	20	16	63	82
Fin. Draw.	114	156	1·14	1·10	72	48	70	85
Spinning	78	78	0·056	0·056	1,800	1,800	90	90

E: existing system *R*: revised system

WASTE IN THE SYSTEM

At each stage in the production line a certain amount of waste is inevitable. This waste may be divided into three sorts: (1) Clean re-usable waste such as sliver ends, thread ends, and bale ropes; (2) dirty waste containing a certain amount of re-usable fibre, this type of waste is found under machines and in floor-sweepings and is passed through the dust shaker to recover the short fibre which can then be put into a sacking weft batch or equivalent; (3) true waste, i.e. mill dust, stick, and other fibre trash which is of no use in the mill.

When the bales are opened at the start of the production line some 4–5 lb of bale ropes represent the first loss, followed by about 1 per cent of dust and stick as the bales pass through the opener. At the softener or spreader roughly ½ per cent of the weight fed falls beneath the machine. At the cards the droppings usually amount to 1½ per cent at the breaker and ¾ per cent at the finisher. Besides the card droppings, there will be in the region of 2 per cent clean sliver waste. Little is lost over the drawing frames—over the complete drawing system about 1 per cent will be lost, while at spinning about another 1 per cent is the normal figure. Drawings and spinning will usually lead to about 1 per cent of clean sliver waste. Much of the waste is re-usable in lower grade batches and even the droppings from the mill will yield about 60 per cent of re-usable fibre from the dust shaker.

The factors which influence the quantities of waste in the mill are:

(1) Good housekeeping. A tidy mill with good cleaning schedules, both for the machinery and the buildings themselves will generally have less waste than a slovenly kept mill.

(2) Fibre quality. At all times a lower grade of fibre will produce more dust and waste than a good grade.

(3) Machine loading. Heavily loaded machines produce more waste than properly loaded ones, particularly in the quantities of sliver waste they produce as a result of choking and lapping.

(4) Oil content. Material processed with low oil contents, such as 'stainless' jute, has a higher waste figure than jute with the higher oil contents. The waste figures increase, particularly at the cards, when the oil content falls below about 3 per cent.

(5) Moisture regain. Dry slivers tend to make more fine dust than those with the proper regain.

(6) Machine speeds and settings. More waste is produced from machinery run at a higher rate than normal, and poorly adjusted machines likewise lead to higher waste figures. At the cards there is an indication that closer settings of the shell, workers, and strippers to the main cylinder bring about a rise in the amount of waste produced.

MATERIAL BALANCE

Because of the waste produced as the material progresses over the system it is inevitable that the weight of bone-dry fibre present in the yarn for sale is less than that in the bales of jute purchased. But since jute spinning on a commercial scale is impracticable without the addition of oil and water this introduces another factor which must be considered when a balance is made between the weight of material fed at the start of the process and the weight of material available for sale at the end of it.

As the batched jute passes over the system the moisture regain steadily falls until, by the time it has been spun into yarn, almost all the water that was added at batching has been lost. The moisture is lost chiefly through evaporation, small amounts being lost in the droppings beneath the machines. Evaporation is greatest at those points where the fibres are exposed to high-speed air-currents, viz., carding and spinning. Inside the cards the fibre on the cylinder travels at roughly 35 m.p.h. and on the spinning frame it is exposed to draughts as high as 15 m.p.h. The quantity of moisture that is lost depends upon the regain of the sliver initially and the relative humidity of the surrounding atmosphere. Table 10.9 gives the results of a series of tests carried out to investigate the effects of these two variables during spinning.

TABLE 10.9. MOISTURE LOSS AT SPINNING

R. H. at spinning (per cent)	*Yarn moisture regain (per cent)*			
	Sliver moisture regain (per cent)			
	12·4	16·1	19·7	22·3
40	10·9	11·8	13·5	14·2
50	11·8	12·9	16·3	16·6
60	12·1	14·4	16·6	17·7
70	14·7	15·4	17·7	17·3
80	15·0	—	20·2	21·5

In addition to this atmospheric effect there are different losses when yarn of various counts is being spun, e.g. for a 12 lb/sp yarn the percentage change in regain is only $\frac{2}{3}$ that for 8 lb/sp yarn. Part of this reduced loss is due to the fact that the heavier yarn is spun with less twist and consequently the yarn is exposed to the atmosphere for a shorter time.

The following moisture regains are found in systems for hessian yarns:

Spreader sliver	33 per cent
Breaker card sliver	29
Finisher card sliver	27
Finisher drawing sliver	26
Yarn (on bobbin)	19

For sacking yarns containing root cuttings it is necessary to apply around 30 per cent of emulsion to the cuttings in order to initiate heating in the pile and under these circumstances the regain at the beginning of the process is rather higher than with hessian qualities.

The quantity of batching oil which is lost varies from mill to mill but it is generally of the order of 10 per cent of the amount added.

The term 'yield' is used in calculations of the amount of yarn produced from a certain quantity of raw jute. If 1 ton of raw jute is brought from the warehouse into the mill and processed in the usual manner and from it 0·98 tons of yarn are available for sale, then the yield is said to be 98 per cent, if 0·96 tons of yarn are made then the yield would be 96 per cent and so on

$$\text{yield} = \frac{\text{wt of yarn produced} \times 100}{\text{wt of raw jute used}} \text{ per cent}$$

In calculations of yield three factors should be taken into account, viz., fibre loss, moisture loss, and oil loss. Generally, only gross losses are considered, that is to say the combined loss of fibre plus oil plus moisture. This simplifies the calculations and under normal circumstances is sufficient, but if an investigation into yield is being carried out then it is better to evaluate the losses separately.

The form of the yield calculations can be expressed quite simply.

The yield will be more than 100 per cent if

(1) more oil is applied than fibre is lost and the raw jute and yarn regains are equal;

(2) the same amount of oil is present as fibre is lost but the yarn regain is higher than that of the raw jute;

(3) more oil is present than fibre is lost and the yarn regain is higher than the raw jute regain.

The yield will be 100 per cent if,

(1) the oil in the yarn equals the fibre loss and the raw jute and yarn regains are the same.

The yield will be less than 100 per cent if,

(1) less oil is added than fibre is lost and the raw jute and yarn regains are equal;

(2) the oil added equals the fibre loss but there is less moisture present in the yarn than there was in the raw jute;

(3) less oil is added than fibre is lost and the yarn regain is lower than the raw jute regain.

An illustrative example of a full material balance is given below, taking account of oil, water, and fibre losses separately. For routine purposes this could be simplified but it is considered that the essentials are:

(1) The weight of bales opened.
(2) The weight of oil used.
(3) The weight of yarn spun and its moisture regain.
(4) The weight of waste collected.

Material Balance

Raw jute	
Bales used	152 = 60,800 lb
Bale ropes collected	760 lb
Nett wt into process	60,040 lb
Moisture regain of raw jute	15%
Dry fibre into process	52,100 lb
Moisture into process	7,840 lb

Emulsion	*Water*	*Oil*	*Emulsifier*
Initial meter reading	37,263	1,539	639
Final meter reading	38,148	1,882	656
Gallons used	875	343	17
lb per gal	10	8·9	10
lb used	8,750	3,050	170
Total emulsion mixed		11,970 lb	
Percentage application $\left(\frac{11{,}970 \times 100}{60{,}040}\right)$		19·9	
Percentage oil on dry fibre weight		5·9	
Percentage emulsifier on dry fibre weight		0·3	
Percentage water on dry fibre weight		14·5	

Yarn	
Total wt of yarn spun	61,250 lb
Yield $\left(\frac{61{,}250 \times 100}{60{,}800}\right)$	101·6%
Moisture regain	19%
Oil content on dry wt of fibre	5·5%
Total wt of fibre	49,200 lb
Total wt of oil	2,710 lb
Total wt of moisture	9,350 lb

Material balance	*Into process (lb)*	*Out of process (lb)*	*Loss (lb)*	*Percentage loss*
Fibre	52,100	49,200	2,900	5·5
Oil	3,050	2,710	340	11·1
Moisture–raw jute	7,840			
emulsion	8,750			
	16,590	9,350	7,240	43·6
Total	60,800	61,250	10,480	

Waste	
Dust from shaker	1,900 lb
Re-workable waste	1,850 lb
Bale ropes	760 lb
Total re-workable waste	2,610 lb
Percentage re-workable waste $\left(\frac{2{,}610 \times 100}{60{,}800}\right)$	4·3

PRODUCTION COSTS THROUGHOUT THE SYSTEM

In all systems, whether they are hessian or sacking, the production costs increase as the material progresses through the manufacturing stages. To illustrate this, two items of the running costs have been selected, the power and labour requirements. The latter will vary from mill to mill and country to country but the figures in Table 10.10 have been found to be representative of efficient operation in hessian systems. All the charges have been expressed on the basis of 100 lb of yarn spun.

TABLE 10.10

	Operative hours	*Kilowatt-hours*
Spreader	0·14	0·46
Carding	0·28	2·70
Drawing	0·50	1·90
Spinning	1·00	13·0

It can be seen that spinning is by far the most costly stage as far as power and direct labour costs are concerned, the same applies to other charges such as depreciation, floor-area costs, and lighting.

PRODUCTION ESTIMATES

Jute machinery is particularly suited to the demands of high output; but maximum production can only be achieved under certain circumstances. One of the factors leading to a sub-capacity output is the diversity of counts and qualities that are produced; this certainly makes the task of the production planning department and the mill supervisory personnel more arduous. The more often a machine has to be changed from one quality to another the greater is the lost time and the more chance there is of inadvertent mixing of qualities—this latter point is of particular importance when low oil content and high oil content material are being produced simultaneously. However, it is always necessary to be able to fulfil sales requirements and where these demand relatively short runs on one count or quality, the supervisory system must be sufficiently flexible to cope with them.

In this respect, it is vital to have a reliable estimate of the production capabilities of each stage in the process, the waste and moisture losses, and so on. As an example, one method of arriving at the number of spinning spindles required to meet certain demands will be shown. The first step is to set up the outputs at 100 per cent from one spindle for the range of counts that the mill spins. An example is given in Table 10.11.

Having determined the 100 per cent spindle outputs for the range of counts then the particular numbers of spindles per frame and the operating efficiency of the mill is used to adjust these units to more

TABLE 10.11

Flyer (r.p.m.)	*Count*	*Twist*		*Delivery speed (yd/hr)*		*100% production spindle/hr*	
		Warp	*Weft*	*Warp*	*Weft*	*Warp*	*Weft*
3,900	$7\frac{1}{2}$	4·3	4·0	1,510	1,625	0·787 lb	0·847 lb
	8	4·0	3·8	1,625	1,715	0·903	0·955
	$8\frac{1}{2}$	3·9	3·6	1,670	1,810	0·985	1·070
3,750	9	3·8	3·5	1,645	1,785	1·025	1·115
	$9\frac{1}{2}$	3·7	3·4	1,695	1,835	1·110	1·210
	10	3·6	3·3	1,735	1,900	1·205	1·320

practical ones. For example, if all the frames had 100 spindles, the average spinning efficiency was 90 per cent and an 8 hr day is worked, then the daily frame outputs can be found:

$$100\% \text{ eff. output} \times 100 \times 8 \times \frac{90}{100} \text{ lb}$$

Count	*Output/frame/day*	
	Warp	*Weft*
$7\frac{1}{2}$	566 lb	610 lb
8	650	687
$8\frac{1}{2}$	709	770
9	738	804
$9\frac{1}{2}$	800	871
10	866	950

These production constants are then ready for use, e.g. how long will it take to spin 5,000 lb of $8\frac{1}{2}$ lb/sp weft if 2 frames are allocated to the order?

Output of $8\frac{1}{2}$ lb/sp weft = 700 lb/frame/day

$$\text{Time to fulfil order} = \frac{5{,}000}{770 \times 2}$$

$$= 3{\cdot}25 \text{ days}$$

How many frames must be allocated so that 20 tons of $9\frac{1}{2}$ lb/sp warp will be produced in 7 days?

$$\text{Output of } 9\tfrac{1}{2}\text{ lb/sp warp} = 800\text{ lb/frame/day}$$

$$\text{Number of frames required} = \frac{20 \times 2{,}240}{800 \times 7}$$

$$= 8$$

In exactly the same manner the production constants for each stage in the process can be found so that the time required to meet certain production demands can be found quickly and easily. This is by no means all that is necessary in the way of production planning and programming but it is intended to show one simple approach to the problems of planning machine utilization.

CHAPTER ELEVEN

Winding

As WINDING can be regarded as the first stage in weaving preparation, this chapter will deal only with the main points of the operation. After the yarn has been taken off the spinning frame it is transferred to one of three types of package—spools, cones, or cops. Although, as has been mentioned, these form the first preparatory stages for weaving, the winding department comes under the jurisdiction of the spinner. This it does as a matter of convenience. If the spinner sold his output on bobbins then he would require large stocks of empty bobbins to meet his own requirements and to allow for lateness or non-return from his customers. The spinner therefore winds packages which are suitable for direct sale.

The particular type of package on which the yarn will be wound depends on the yarn's end-use. Warp yarn will be wound on spools or cones; weft on cops, spools, or cones.

SPOOL AND CONE WINDING

In this operation the yarn from a number of spinning bobbins is tied head-to-tail to form a long continuous length of yarn which is wound on a wooden or paper centre. Spools are cylindrical and cones are, as their name suggests, conical. Both packages are without flanges and the yarn is built into a stable formation by winding it at a suitable angle. Spools are commonly 8–10 in. across the face and up to 10 in. in diameter. Cones, designed for over-end yarn removal, may be up to 15 in. in diameter with a 10 in. traverse, holding 45 lb of yarn. Cones generally have a taper of about 10 degrees though greater and lesser angles are found.

Two types of spool or cone may be made—open wound or precision wound. The first, and commoner, of these is made on a machine with a driving drum against which the package rotates through surface contact. As the drum has a fixed speed, it follows that the yarn winding speed is likewise fixed. The yarn may be traversed by means of guides set in a traverse-bar running along the machine; the bar being moved to and fro by a cam. Alternatively, the yarn may run in a helical groove cut in the driving drum itself, the yarn being led through the groove and traversed and wound by the one drum motion.

In drawn-winding the driving principle is straightforward and as the spool diameter increases, the spool r.p.m. decreases to give a constant surface speed. A cone, however, with its varying diameters does not behave in such a simple manner. There is only one point on the cone where the surface speed equals that of the drum. Towards the nose the drum travels faster than the cone and towards the base the cone surface speed is higher than the drum's, see Figure 11.1.

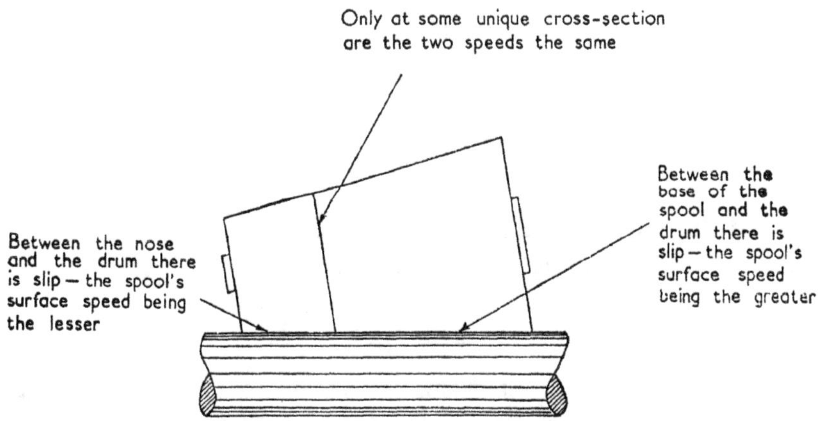

Figure 11.1. Slip encountered when driving a conical spool on a drum-type machine

Since the surface speed of the cone is

$$S = 2\pi fr$$

where f is the cone's rate of revolution and r is its radius, it follows that the surface speed varies from the nose to the base of the package and can only equal the driving drum's at one particular point. Some slip must therefore occur between the cone and the drum.

Open-wound packages have the yarn laid in a relatively open manner, successive layers criss-crossing with the previous in an irregular pattern. Precision spools or cones do not show this irregularity but have the yarns laid contiguously leaving very little free air-space between them. This leads to a very hard, dense package.

Precision winding is achieved by laying a definite number of spirals of yarn on in one traverse of the guide. Precision packages are wound on a machine which has a driven spindle on which the wooden or paper centre is mounted. The traverse guide is driven from the spindle at a certain 'winding ratio', i.e. the number of spindle revolutions to one traverse. If the winding ratio is 3 then three complete spirals will be wrapped round the package in each guide traverse. Note that if the winding ratio is a whole number or a half-number successive layers would be built exactly on top of previous ones and the fault known as ribboning would arise. The yarns would simply build up a spiral band bearing no resemblance to the desired product. To avoid this a slight lead, or gain, is given to the traverse cam-drive so that the yarn is laid in the fashion shown in Figure 11.2.

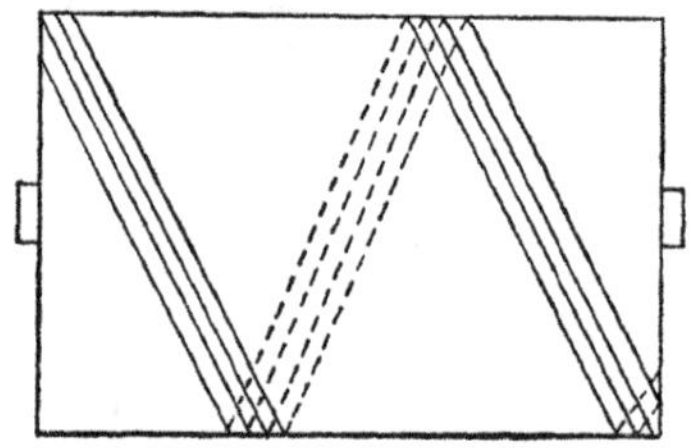

Figure 11.2. Precision-wound spool

COP-WINDING

For flat looms cops vary in diameter from $1\frac{1}{2}$ to 2 in. and in length from 10 to 12 in. while for circular looms they measure up to $3\frac{1}{8}$ in. in diameter and $17\frac{3}{4}$ in. in length. The cop is formed by winding the yarn on a bare spindle which is then withdrawn when the desired length of cop has been wound.

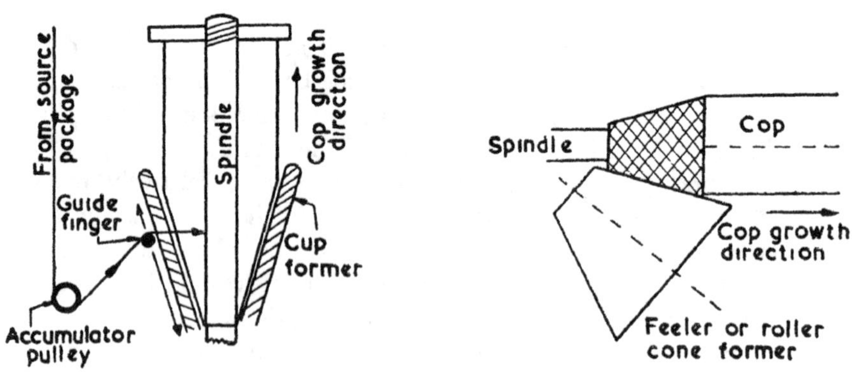

Figure 11.3. Cop-winding methods

The unit of the cop machine consists basically of a rotating spindle, a traverse guide and a nose-forming roller or cup, Figure 11.3.

The guide traverses back and forth, laying the yarn round the spindle in a wide spiral—winding ratios of 1·6–4·8 are found. At the start of cop formation the yarn builds a miniature open-wound spool until it comes in contact with the nose-forming roller or cup. The taper of the nose-forming member decides the nose angle on the cop, for jute yarn this is usually between 15 and 18 degrees for flat loom cops and about 22 degrees for circular loom cops. The cop continues to grow until a stop-motion, set to produce the required cop length, stops the spindle. The cop is doffed by withdrawing the spindle.

To increase the diameter of the cop a longer traverse stroke is used and *vice versa*. To give good unwinding the nose length is made at least ⅛ in. greater than the diameter.

Depending on the design of the machine the spindles may be mounted horizontally or vertically. Modern machines run at 1,500–3,000 r.p.m. and have automatic doffing and re-starting. To capitalize on the high efficiencies of modern cop machines, the supply yarn is either on spools, cones, or tag-end bobbins, i.e. bobbins in which the first end on the bobbin is led to the outside of the full bobbin so that it may be tied to the tail-end of the next. Older machines have vertical hand-doffed spindles running at 1,000 r.p.m. and are fed from individual bobbins.

PRODUCTION ASPECTS OF WINDING

There are between 500 and 1,000 yd of yarn on a spinning bobbin. This means that the winder must tie at least one knot every 500–1,000 yd (there will be others to repair breaks but these are relatively infrequent and will be ignored for simplicity's sake as there are only about 2 per 10,000 yd of yarn). The winding machine's output is closely related to the winder's work-load, which in turn is intimately bound up with tying knots between bobbins.

Consider a drum-type spool machine running at 180 yd/min with a fresh bobbin waiting on each of its 40 spindles. If the bobbins hold 900 yd of yarn, each will be wound in 5 min. The winder starts 1, moves to 2 and starts it, moves to 3, and so on. If the winder can start bobbins at 10 sec intervals then it takes

$$40 \times 10 = 400 \text{ sec } (6{\cdot}67 \text{ min})$$

to work along the whole machine. By this time bobbin 1 will have wound, run out, and been idle for 100 sec (400 − 300), bobbin 2 has been idle for 90 sec (400 − 310), bobbin 3 for 80 sec (400 − 320), etc., up to bobbin 11 which is just finishing as bobbin 40 begins. The total lost time from bobbins 1 to 11 is 550 sec (9·18 min).

If now the speed of the machine is increased to 300 yd/min each bobbin will be wound in 3 min. A similar calculation shows that the lost time is now 2,330 sec (38·8 min). In both cases the net winding speed is

$$\frac{40 \times 900}{6{\cdot}68} = 135 \text{ yd/min}$$

At the slower speed this corresponds to an efficiency of 76 per cent but at the higher speed the efficiency is only 45 per cent. It must be remembered that efficiency or winding speed in themselves mean little —it is the combination of them both which determines output.

This presents a very simple picture of the more complicated practical case but should be sufficient to show that machine efficiencies, machine speeds, and output must be studied carefully if successful operation is to be achieved.

Just as material flow played its part in earlier operations the numbers of winding spindles, their speeds, and their efficiencies must be related to the spinning production.

For example, a mill has 20 spinning frames on 8 lb warp (4,000 r.p.m., 4 t.p.i.), 10 frames on 8 lb weft (3,900 r.p.m., 3·8 t.p.i.), 12 frames on 10 lb weft (3,700 r.p.m., 3·6 t.p.i.) and 4 on 11 lb weft (3,700 r.p.m., 3·5 t.p.i.); all the frames have 100 spindles. Spools are wound at 210 yd/min at 70 per cent efficiency, cops at 50 yd/min at 80 per cent from spools. How many warp and weft spindles are required?

Spin output at 90 per cent efficiency.

$$\text{Warp:} \quad \frac{20 \times 4000 \times 90 \times 100}{100 \times 4 \times 36} = 50000 \text{ yd/min}$$

$$\text{8's weft:} \quad \frac{10 \times 3900 \times 90 \times 100}{100 \times 3{\cdot}8 \times 36} = 25600$$

$$\text{10's weft:} \quad \frac{12 \times 3700 \times 90 \times 100}{100 \times 3{\cdot}6 \times 36} = 30850$$

$$\text{11's weft:} \quad \frac{4 \times 3700 \times 90 \times 100}{100 \times 3{\cdot}5 \times 36} = 10570$$

$$\text{Total:} \quad \underline{117020} \text{ yd/min}$$

All this yarn must be wound into spools at an effective speed of

$$\frac{210 \times 70}{100} = 147 \text{ yd/min}$$

Spooling spindles required

$$\frac{117020}{147} = 795$$

Total weft spin output 67,020 yd/min.

Effective weft winding speed

$$\frac{50 \times 80}{100} = 40 \text{ yd/min}$$

Cop spindles required

$$\frac{67020}{40} = 1676$$

WINDING FAULTS

A not uncommon difficulty with cops is wrong diameter or variations in diameter. For automatic weaving it is particularly important that cops should be of the correct size otherwise faulty loading will result at the loom. Short-nosed cops can cause loops and snarls in the cloth when the cop breaks down in the shuttle. If the moisture regain varies widely then irregular cloth widths will result, for experience shows that cops with high moisture regains give narrower cloth than those with lower regains.

Spools may exhibit 'cobwebbing', Figure 11.4, where the yarn has passed over the end of the spool.

Figure 11.4. End view of a 'cob-webbed' spool

When these spools are unwound the yarn breaks as each trailing cobweb jerks the spool. If this is occurring on all spools from one bank of spindles it indicates a worn cam or badly positioned traverse bar. On individual spools the trouble may be due to a worn or slack guide or wrong positioning of the spool relative to the traverse guide.

TWIST CHANGES AT WINDING

If yarn is drawn off the side of a bobbin, spool, or cone no twist change takes place, but if it is drawn over-end from these packages or from a cop then the twist per unit length in the yarn is changed. The size of the change depends on the length of yarn in each spiral on the package. For instance, if one complete spiral of yarn on a spool is 12 in. long then over-end removal will change the yarn twist by 1 turn in 12 in., i.e. 0·083 t.p.i. Notice that the shorter the length of the spiral the greater is the twist change. Thus in a cop where the length of each complete spiral of yarn may be only $3\frac{1}{2}$ in. the twist change would be 0·29 t.p.i. The general level of twist change in cops is about 0·25 t.p.i.

The direction of unwinding, clockwise or anticlockwise, determines whether the twist change will be positive or negative, i.e. whether extra twist will be put into the yarn or taken out of the yarn. As almost all jute yarns are spun with Z-twist only this case will be considered (if S-twist yarn is used then the effects are the opposite, gains become losses and *vice versa*). If, when viewed from the end over which the yarn is drawn, the yarn moves in an anticlockwise direction then twist is taken out. If, on the other hand, the yarn rotates clockwise then twist is added.

CHAPTER TWELVE

Quality Control

As THOSE industries which use jute undergo economic and technological change their emphasis on reliable yarn quality becomes more insistent. At all times there is a demand for either a better product at the same price or a yarn equivalent to current standards but costing less. In general, quality levels in jute goods are set by the markets in which they are sold and have been evolved over the years through normal commercial usage. But whatever standards are set by the market, the producer must have his own standards of quality. Commercial and production standards should not be far apart for if the quality is lower than the customer will accept there will be loss of markets but if it is much higher than necessary then production costs will be unnecessarily high. To many people, quality control is synonymous with testing carried out by a special department, but this is not so. Quality is decided on the shop-floor by the grade of fibre that is used and the effectiveness with which it is processed. Testing will never control quality, it will merely indicate when the product is off-standard and it is the responsibility of the production staff to maintain quality levels.

Since the quality standards and control requirements vary from mill to mill it is impossible to formulate one scheme which will suit every case, therefore in this chapter a few methods of process control will be discussed, with particular reference to those which can be carried out on the shop-floor. However, it is worth noting that before any scheme can be initiated it is essential that:

(1) The testing methods are sound and designed to yield information that can be acted upon quickly.
(2) Shop-floor tests and record keeping are as straightforward as possible and do not interfere greatly with the manufacturing process.
(3) There is a person in the organization who can interpret the test results and recommend certain courses of action.
(4) There is a genuine desire to achieve a good degree of control over the process and a determination to succeed.

(5) It should be recognized that it may take several years before control is firmly established.

In setting up a control scheme there are five main points which require attention; the manner of using the machinery, the selection of the batch, the amount of moisture and oil present in the product, the count of the slivers and yarns and the running efficiency of the process. Experience with many mills has shown that these factors commonly give rise to trouble and are very often the principal points on which a scheme of process control can be begun.

THE MACHINERY AND ITS USE

Sound maintenance is the basis for good performance from the points of view of production and quality. The machinery makers' recommendations should be examined carefully and a routine for cleaning, lubricating, renovating, and replacing should be organized since it is very often more beneficial to have regular short stoppages for maintenance than one or two long stoppages for expensive repairs. The following points give an indication of the type of work which is required.

In the emulsion plant all pumps, filters, and agitators should be stripped and cleaned at regular intervals. If possible replace all sight-glasses with fluid meters but, having done so, it is necessary to subject them to periodic calibration checks to see that they are maintaining their accuracy. At the spreader all the pins on both chains should penetrate the sliver and broken or bent pins should not be tolerated; pressure gauges or flowmeters require regular attention and all filters in the supply line should be cleaned. One part of the spreader which may sometimes be overlooked is the flex-drive to the gear-box of the feed-indicating mechanism; this requires greasing and should operate with as smooth a line from the spreader to the weighbridge as possible—any kinks or sharp bends in the drive will cause the driven pointer to jerk, making it more difficult for the operative to maintain the proper rate of feed.

At the cards it is necessary to examine the state of the pins on all the rollers from time to time; hooked, grooved, or blunt pins indicating that re-staving is required. In poorly maintained cards complete gaps in the pinning may be seen where all the pins have broken away completely. At regular intervals the roller settings and alignment should be checked.

On the drawing frames correct pinning is essential and at no time should a faller-bar be allowed to work with broken or missing pins. The paths and slides which carry the fallers should be clean and unworn, the rollers should be checked regularly for signs of wear and misalignment.

On the spinning frame the rollers should be checked for alignment; all the spindles should be exactly central to the flyers; the tapes driving the flyers should be tight or there will be an excessive amount of slip at the wharf and the flyers may not rotate at the proper speed and the yarn twist will be low; the slide carrying the builder should be clean and the builder should move evenly up and down—a jerky movement causing irregular spinning tension pulses which will increase the end breakage rate. The rubber covers on the drafting pressing rollers should be true to their rollers for if they buckle there is a tendency for the fibres to work out of the nip and cause a yarn break. The drag-pads should be kept clean and where the four-pad type is used, the pads should be in the same position on all carriers—on the inside for low counts, in the middle for medium counts, and on the outside for high counts. The bobbins themselves should be free from rough or jagged edges that might catch in to the yarn and cause a yarn break.

These are but a few of the types of work needed for good maintenance. A log-book, kept over a period of several months, of the reasons for machine stoppages and of the spares issued will provide a basis on which to build a maintenance scheme.

With regard to the operation of the machinery, perhaps the commonest cause of low-quality work is overloading the machine and running it too fast. At the cards, an excessive weight on the feed sheet will result in frequent chokes and laps, the card efficiency will fall, and proper carding will not be carried out. Sufficient has been said about correct pinning on the drawing frames being essential for good quality work. High speeds not only lead to more wear and tear on the faller-bars but prevent the sliver from being pinned properly. On the spinning frame, excessively high speeds result in greater numbers of yarn breaks and cause the yarn to be 'hairier' than normal (though this last feature depends on twist, fibre quality, moisture regain, etc.).

BATCH SELECTION

The prime factor in setting the level of the batch is the price. It is one of the axioms of jute spinning that the cost of the batch will be as low as is consistent with quality standards. One must examine critically, however, the processing potential of the blend of fibre for unless the material will process without excessive difficulty any price advantage gained from using a cheap batch will be lost in the extra processing costs. The amount of waste which arises at the various stages must be considered too, for this has to be reworked and should be debited against the production line from which it arises. (Incidentally, it is often illuminating to examine those stages in the process where large amounts of waste accumulate—this is sometimes a sign that the machines are poorly maintained or severely overloaded.) Fibre quality has a great bearing on the appearance of the yarn and its strength characteristics. Low grade fibre increases the degree of short-term irregularity (the 'thicks and thins') and produces a weaker yarn. To show this effect, the results of a series of tests, in which pure unblended strains of fibres were spun into 276 tex yarn and tested, are given in Table 12.1.

TABLE 12.1. EFFECT OF FIBRE GRADE ON YARN QUALITY

	Mill Lightnings	*Premium Hearts*	*Mill Hearts*	*Grade Hearts*
Relative price at time of test	100	90	87	69
Fibre diameter (microns)	37	—	—	40
Trash content (per cent)	5	11	13	30
Short term irregularity (per cent)	30	33	39	42
Tenacity (g/tex)	11·9	12·1	10·7	8·9
C.V. of breaking load (per cent)	23	24	26	27

Trash content—Amount of extraneous matter, bark, root, stick, etc., present at the breaker card.

C.V. breaking load—A measure of the spread of the breaking load results, the higher the C.V. the poorer the yarn.

Short term irregularity—Measured on ½ in. lengths of yarn.

Clearly, the grade of fibre chosen has a direct bearing on the quality of the yarn produced—the processing machinery used can only operate on the fibre presented to it and in this way the general quality level is 'built in' at the batch.

MOISTURE AND OIL LEVELS IN THE MATERIAL

Commercially, it is important to dispatch yarn with the correct quantity of oil and water present in order that the maximum profitability may be achieved. Technically, the moisture regain has a great bearing on carding, drawing, and spinning while the oil content of the yarn must meet the end-use requirements. The first control point for moisture and oil is at emulsion preparation. The operative responsible for mixing the emulsions should have clear instructions for the amounts of each ingredient and the method of mixing. In addition, there should be an adequate supply of metering units, gallon and pint measures, scales, etc. The simplest method of checking that the emulsion is being made correctly is to 'crack' a sample, i.e. deliberately break the emulsion so that it separates into two phases which can then be measured.

Method 1—suitable for all types of oil-in-water emulsions. In this test a definite volume of emulsion is cracked with acidified sodium sulphite and the separated oil is measured. A sample of emulsion is drawn off, preferable from the sprays or the weir, some 110 ml being a suitable sample size for routine purposes. The sample bottle is shaken well and 100 ml measured from it into a measuring cylinder and then transferred to a beaker and heated to 90–95° C. 10 ml of 10 per cent sulphuric acid and 5 g of anhydrous sodium sulphite are added to the measuring cylinder and the hot emulsion poured back into it. The contents of the cylinder are stirred well with a glass rod and the oil allowed to separate into an upper layer and its volume measured; if there are v ml of oil in the top layer then the emulsion contains v per cent oil and $(100-v)$ per cent water. After the hot emulsion has been put back into the cylinder never shake or invert the contents since the rapid evolution of gas may force some of the hot acidic solution out of the vessel.

Method 2—suitable for emulsions prepared with ionic or soap-type emulsifiers only. From a well shaken sample of about 110 ml, 100 ml are measured off into a measuring cylinder and 10 g of common salt (or 10 ml of 10 per cent sulphuric acid) is added and the contents shaken and allowed to settle. Again the oil forms an upper layer the volume of which is read off and the oil content calculated in the same way as Method 1. It may help the emulsion to break if the sample is warmed slightly.

Having confirmed that the correct emulsion recipe has been adhered to, the next step is to ensure that the correct amount of emulsion is being applied to the jute. In this respect, flowmeters are especially useful since they show at a glance the emulsion flow rate and it is a matter of a short calculation to arrive at the percentage application rate. There is, however, another method by which the percentage application of emulsion can be found, viz. 'add-on' tests, in which the weight of dry jute fed and the weight of batched jute are recorded, the difference being the 'add-on' of emulsion.

One very useful instrument for monitoring the oil content of the material in process is the ultra-violet lamp. The mineral oils used for jute batching possess the property of fluorescing under ultra-violet light and the amount of fluorescence present depends on the quantity of oil. Thus, slivers with 5 per cent oil show a much stronger violet glow than those with only 1 per cent. This immediately gives a valuable means of distinguishing between normal oil yarn and 'stainless' yarn, indeed, with experience and on the same colour of fibre, two samples can be distinguished from each other even if their oil contents are only 0·5 per cent different. The lamp should be used regularly to see that no 5 per cent oil yarn or sliver becomes mixed with 'stainless' material. Defects in 'stainless' goods arising from oil stains caused by careless use of an oil-can or from oily caddis from the machines falling on to the low oil content jute can be seen easily under the U.V. lamp. Certain highly refined oils, such as 'Odimin' oil do not fluoresce under U.V. and therefore could not be distinguished from 'stainless' material but, in these cases a special fluorescent substance can be put into the emulsion which will show up under the lamp and permit the 'Odimin' jute to be identified. Only extremely small amounts of these tracers are required and they can be obtained to fluoresce in yellow or green to differentiate between the 'Odimin' jute and the normal 5 per cent material.

With the development of electronic moisture meters, the measurement of moisture regain has become much simpler and indeed it may be said that without them it was impossible to sample sufficient material and test it quickly enough to provide an effective means of process control. The B.J.T.R.A. Probe Moisture Meter, Plate VIII, can be used to measure the moisture regain of raw jute, spreader rolls, and card rolls, and other types, such as the Shirley Moisture Meter or the Marconi Moisture Meter can be used for bobbins. The basis for the quantity of moisture in the material is laid at the spreader or the softener and the remarks already made about the emulsion content and application apply equally well to the moisture regain as to the oil content. Before the moisture levels can be brought under control the technique of mixing and correct application must be firmly established. When the jute is in the production line, tests may be made with an electronic meter at each stage but it is usually sufficient to leave such checks until the yarn has been spun. If it is found that the moisture regain of the yarn has changed then concentrated tests can be carried out throughout the process in order to find the cause. Moisture testing at the yarn stage is essential if correct control over the count is to be achieved.

COUNT CONTROL

When a scheme for quality control is introduced into a mill the factor which causes most difficulty is the variability of count in slivers and yarns. It is almost impossible to draw valid conclusions from small samples or short tests. Wherever it can be done it is better to extend special tests over a period of several weeks or even months and to collect small amounts of information at random intervals during that period and consider them as a whole before passing judgement. As far as day-to-day testing is concerned it is essential that statistical control charts be used so that premature action will not be taken on apparently high or low results which arise solely from the natural variations in count. It is usually found that after control charts have been used for some time the number of changes which are made to sliver weights and draft pinions are markedly reduced. (The reader who is unfamiliar with the methods of compiling and using these charts is referred to the 'Further Reading' list at the end of this book.)

Testing the count of jute slivers or yarns is complicated by the presence of variable quantities of moisture—a yarn may be below count or above count simply because of moisture. If accurate count testing is to be done, it is essential that the moisture regain of the material under test is known. Fortunately, with the modern moisture meters available this is comparatively simple. Unless the moisture regain is tested simultaneously with the count the conclusion may be drawn that the count is heavy or light solely because the moisture regain is varying. There are practical difficulties in using some types of moisture meter on finisher drawing sliver but since the main count control point is at the spinning frame this is not a serious disadvantage.

The object of controlling the count at the spinning frame is to produce a yarn with the correct quantity of *fibre* in it, and at no time should the draft be changed to take account of variations in count due to moisture changes. When the count and the moisture regain have been measured the count should be converted to a standard moisture regain. This is commonly 14 per cent (the desorption moisture regain at 65 per cent R.H. and 20° C). Figure 12.1 shows a control chart for

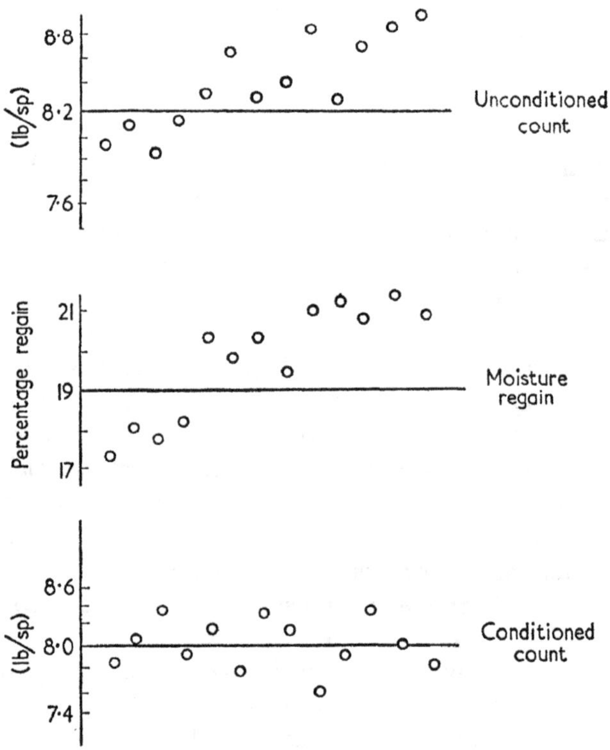

Figure 12.1. Yarn count control charts

yarn count with the uncorrected count, i.e. the count before allowing for the moisture present in the yarn, the moisture regain, and the count corrected to a regain of 14 per cent. The 'uncorrected' chart gives the impression that the yarn is becoming heavier and a draft pinion change is necessary, whereas the real cause of the drift is a change in the moisture regain; if the corrected count chart is examined it will be seen that the yarn count actually remained reasonably constant over the period.

The moisture level at the spinning frame should be such that the yarn leaves the mill with a regain of 13–14 per cent. This requires that the spinning regain at the frame is of the order of 18 per cent to allow for small losses in winding and storage.

TABLE 12.2. SOURCES OF YARN WEIGHT IRREGULARITY

Machine	*Cause of irregularities*	*Effect in yarn*
Spreader	Human error at feed Variations in strick weight Drafting waves Variations in emulsion flow	Long term drifts in count Responsible for 75 per cent of bobbin-to-bobbin variation
Cards	Variations in feed slivers Gulping Missing doublings	Bobbin-to-bobbin variations in count in small samples Responsible for 25 per cent of between-bobbin count fluctuations
Drawings	Variations in feed slivers Missing doublings Faller-bar slubs Bad splices	1st responsible for variations in count of adjacent 100 yd lengths 2nd for adjacent 20 yd lengths and sets pattern for 'thicks and thins' Finisher accentuates pattern for 'thicks and thins'
Spinning frames	Variations in feed sliver Bad splices in sliver and yarn Incomplete draft control	Sets final degree of short-term irregularity

In Table 12.2 a brief summary of the sources of irregularity in jute slivers and yarns is given in which those factors that influence the gross variations in count are differentiated from those that affect the short-term variations. In general the uniformity of the yarn which is seen when a short length is examined is decided by the regularity of the sliver being fed to the finisher drawing and the spinning frame and the efficiency with which these frames control short fibre movement during drafting. The variations which give rise to changes in weight of long lengths of yarn arise much further back in the process.

PROCESS OBSERVATION

In the mill the supervisory staff are continually making assessments of how well the material is processing by subjective judgements. Over the years they have built up experience of what constitutes good or bad machine performance and their evaluations are based on this experience. In process observation the same methods are adopted but objective measurements are made which are then analysed in a logical manner and used to assess the performance of the material, machinery, or the operatives.

At the spreader, cards, and drawing frames, observations of the number of laps or chokes occurring, the frequency of missing doublings, and waste losses may be made. Such tests can be carried out by continuous observation where the number of defects in a certain time is counted and related to some common basis, e.g. laps per 100 lb of sliver, chokes per hour, etc. Alternatively, random observations may be made in which the number of occurrences on which the fault was seen is compared with the total number of observations made. In this method it is important that the observations be random in time and cover the full work-period and that the observations are made instantaneously. For example, a record of the number of laps at a card may be required. The observer will pass the card at random times over a period of perhaps a week or a fortnight, noting whether the card is running or stopped and, if it is stopped, whether a lap has occurred. The record may appear like this:

Total number of observations	3,370
Number of times card stopped	674
Number of times card stopped for lap	220

From this it may be calculated that

$$\text{Card efficiency} = \frac{\text{Number of times observed running}}{\text{Total number of observations}}$$

$$= \frac{(3370-674)\times 100}{3370}$$

$$= 80 \text{ per cent}$$

$$\text{Percentage of stops due to laps} = \frac{220}{674}\times 100$$

$$= 32{\cdot}6$$

At the spinning frames, process observation is usually limited to counting the number of end breaks. This may be done by continuous observation over a long or short period or by random observations. Continuous observation over a long period is the method usually resorted to when comparisons between two types of yarn or processing methods are being made. The observer stands at the frame and counts the number of ends which fall in a given time. If the yarns do not differ in quality markedly this method cannot yield useful results unless several hours are spent observing each yarn type. For this reason it is time-consuming and tedious. For routine purposes continuous observation over short periods may be carried out. In this method of test the observer spends only 5 or 10 min at each frame and counts the number of breaks. In this way many more frames can be dealt with but there is the complication that the diameter of the bobbin has the usual effect on spinning breaks and when this method is used the approximate diameter of the bobbins should be recorded. The results from frames working on the same quality and count of yarn should then be averaged. In both these methods one must expect quite large variations in the number of end breaks from doff to doff. It has been said that the number of ends falling should be counted, but in practice it is easier and less confusing to count the number of ends that the spinner *repairs*, and counting the number of ends which are idle at the start of the observation period and at the end. For example,

Ends down at the start	6
Ends repaired	35
Ends down at the finish	3
Total number of ends down	35+(6−3)
	= 38

The results of tests of this nature can be expressed in terms of the number of breaks per 100 spindles per hour or of the number of breaks per 1,000 yd or per pound of yarn or some such suitable unit.

If the random method of observation is adopted the observer passes each frame and notes how many spindles are standing idle. After several patrols covering the working period the results are averaged for frames working on the same type of yarn and the average number of idle spindles per frame can be found for that quality. Where the flat contains frames with different numbers of spindles a comparison may be made by calculating the percentage of idle spindles. It is important to realize the fundamental difference between the continuous and random methods of observation. The former shows how the yarn is behaving on the machinery and the latter shows how the spinner is reacting to a certain breakage frequency. The random test is simpler and cheaper to carry out but includes, to a very great extent, the human element and for this reason must be handled circumspectly.

While it is true to say that each mill has a certain degree of quality and process control—otherwise it could not remain an effective production unit—it is equally true that an organized, logical approach to the problem will go far towards solving the difficulties associated with maintaining quality standards in large scale production.

FURTHER READING

URQUHART, A. R., and HOWITT, F. O. *The Structure of Textile Fibres* (Textile Institute)

MATTHEWS, A. *Textile Fibres* (Chapman and Hall)

KUNDU, B. C., BASAK, K. C., and SARCAR, P. B. *Jute in India* (Indian Central Jute Committee)

Indian Jute Atlas (Indian Central Jute Committee)

Marketing of Jute in Pakistan (Dacca University Socio-economic Research Board)

KIRBY, R. H. *Vegetable Fibres* (Leonard Hill/Interscience)

MORTON, W. E., and WRAY, G. R. *Introduction to the Study of Spinning* (Longmans)

MORONEY, M. J. *Facts from Figures* (Pelican Books)

MURPHY, T., NORRIS, K., and TIPPET, L. *Statistical Methods for Textile Technologists* (Textile Institute)

Index

www.ingramcontent.com/pod-product-compliance
Lightning Source LLC
LaVergne TN
LVHW011521200926
848484LV00001B/102
9780820600949